Masami Nanzyo • Hitoshi Kanno

Inorganic Constituents in Soil

Basics and Visuals

Masami Nanzyo
Graduate School of Agricultural Science
Tohoku University
Sendai, Japan

Hitoshi Kanno
Graduate School of Agricultural Science
Tohoku University
Sendai, Japan

Library of Congress Control Number: 2018951424

Preface

There has been increasing interest in our environments, which consists mainly of air, water, and soil environments. Organisms, including humans, live within and interact with these environments. However, the degree of the interest about soil has not been as high as those for air, water, and organisms because air pollution, changes of air temperature, contamination of water, flooding, shortages of water, and increases and decreases in the abundance and diversity of organisms tend to affect us more directly. Changes in soil are less conspicuous because soil has buffering capabilities, to some extent, against various impacts.

The buffering actions of soils, which are an outstanding aspect of soil, are due to the sorption and release of various materials by soil constituents and to biological activities. For example, the buffering action of soil against acids and bases is primarily due to the surface properties of inorganic and organic constituents. In addition, the biological buffering action of submerged soils can result in the reduction of nitrogen oxides and sulfur oxides added to soils by acid deposition.

Soil may appear very typical or similar when considering only the familiar surface of soil. However, once one has an opportunity to observe a different soil, he or she may become interested in the soil because of its unique aspects. To find a different soil, one needs only to visit a place with a different landscape or dig into soil to a depth of 0.5–1 m. Different soils often exist beneath different landscapes, and as many soils show differentiation into various horizons, there may be differently colored horizons beneath the surface of the soil. These observations lead to questions about the reasons for the differences between typical soil and the newly found soil.

One answer to these questions is the difference in soil constituents between various soils. Soil inorganic constituents are related to brown, red, yellow, black, and white colors of soils, whereas soil organic constituents are related to dark or dark-brown colors. Redox reactions of soil inorganic constituents are also related to blue color and brown mottling in soils. The feel and consistence of soils also depend on the properties and composition of soil constituents.

For our discussions of soil inorganic constituents that provide distinctive properties to soils, we consider noncrystalline materials and constituents sensitive to redox reactions. We also discuss somewhat special events that exist in various places including Japan, such as the effects of tsunamis and radiocesium pollution, because these are related to the functions of soil inorganic constituents.

In this monograph, we introduce and identify the inorganic constituents in soil so that readers can obtain an overview of them quickly. We list references for further study at the end of each chapter. We hope that this monograph will contribute to an understanding of soil, to efficient soil use, to conservation of soil, and to keeping our environments comfortable.

Sendai, Japan
March 2018

Masami Nanzyo
Hitoshi Kanno

Acknowledgments

We thank all those who provided advice and assistance toward completing this monograph.

In relation to Chaps. 1, 2, 3, 4 and 5, many people assisted our soil sampling and analytical work. The late Dr. T. Takahashi, project leader of the Soil Development and Research Center, Japan International Cooperation Agency, provided us an opportunity to survey and collect samples from Central Luzon, Philippines. Dr. Tokutome collaborated with us during the soil sampling in the central plain of Luzon, Philippines. Professor N. Mizuno of Rakuno Gakuen University provided a sample of the Tarumae-a tephra, Hokkaido, Japan. Dr. Jae-Sung Shin, deputy director general of the National Institute of Agricultural Science and Technology, Korea, and his staff provided soil samples from Singun-ri, Republic of Korea. Professor Manuel Casanova of the University of Chile provided Palexeralf sample rich in kaolinite. Professor M. Nakagawa of Kochi University provided a chlorite schist sample. Kunimine Industries Co., Ltd. provided a sample of Kunipia F. Emeritus Professors S. Shoji and M. Saigusa and Dr. I. Yamada of Tohoku University provided precious advice and samples of Andisols in Japan collected through their research projects. Emeritus Professor S. Yamasaki provided analytical data for 57 elements in Andisols in Japan. Professor H. Hirai of Utsunomiya University collaborated with us during soil sampling at Kiwadashima, Tochigi Prefecture, Japan. Emeritus Professor H. Takesako of Meiji University provided soil samples from his study site at Hadano, Kanagawa Prefecture, Japan. Dr. N. Yasuda of the Mie Prefectural Agricultural Experiment Station assisted with Andisol sampling in Suzuka, Mie Prefecture. Professor M. Watanabe provided advice helpful for separating and identifying sclerotia grains. Professors R. A. Dahlgren of the University of California, Davis, and Z. S. Chen of National Taiwan University helped us to study Andisols in California and Taiwan, respectively. Drs. K. Togami and K. Miura of Tohoku Agricultural Research Center cooperated with us in the soil sampling of a paddy field in Miyagi Prefecture. Dr. S. Arai of Mitsubishi Materials Corporation helped with soil sampling in Miyagi Prefecture and partially supported this study. Dr. B. Harms of the Department of Environment and Resource Management, Brisbane, provided a natrojarosite sample.

In relation to Chap. 6, many people, agencies, local governments, and private companies provided support and cooperation for our research projects. These include Professors M. Saito, Y. Nakai, T. Ito, and T. Takahashi and Assistant Professor M. Omura of Tohoku University; farmland owners Mrs. S. Hiratsuka and H. Sasaki; Dr. H. Sugimoto of the Obayashi Corporation Technical Research Institute; Sendai City staff; Miyagi Prefecture staff; the Japan Science and Technology Agency; Asahi Industries Co., Ltd. and its subsidiary; the Naito Foundation; and the Kureha Corporation. Professors T. Nakanishi and K. Tanoi and Dr. N. Kobayashi of the Graduate School of Agriculture and Life Sciences, Tokyo University, contributed to the survey of radioactive particles in sidebar deposits using the imaging plate. Dr. A. Takeda of the Institute for Environmental Sciences of Japan cooperated with us in the imaging plate procedure. Mr. A. Hio of Tohoku University operated the γ-ray spectrometer. Mrs. K. Kurita and T. Yamaguchi of the Japan Broadcasting Corporation collaborated on soil sampling. The National Institute for Materials Science and the Japan Atomic Energy Agency supported the radiocesium studies in Miyagi, Fukushima, and Niigata Prefectures. The former National Institute of Agro-Environmental Sciences, Japan, provided the apatite samples.

Regarding the whole monograph, students and staff of the Soil Science Laboratory, Tohoku University, collaborated on sampling and analyses of soils, and Ms. K. Ito and the late Mr. T. Sato of Tohoku University helped operate the electron microscopes. Dr. Mei Hann Lee of Springer Japan offered fine support and technical assistance that facilitated the publishing of this monograph. This work was partly supported by JSPS KAKENHI Grant Number 17K07693.

Graduate School of Agricultural Science
Tohoku University
Sendai, Japan
March 2018

Masami Nanzyo
Hitoshi Kanno

Contents

Chapter 1
Purpose and Scope

Abstract Soil plays a major role in ecosystem services (an ecological term referring to the benefits granted to humans by ecosystems). Among the various ecosystem services, provisioning services are important providers of foods, fibers, wood, and other naturally sourced materials. To increase biological production while maintaining sustainable soil and ecosystems, we must understand element cycling in soils and ecosystems and its high dependence on inorganic constituents. This monograph describes the fundamentals, along with visual aids, of inorganic soil constituents for beginning students of soil science, environmental science, biogeochemistry, and interested readers in other disciplines. The visual aids include optical photographs, electron microscope images, and element maps acquired by energy dispersive X-ray analyses.

1.1 Ecosystem Services as an Embodiment of Soil Functions

An ecosystem is characterized by a wide variety of organisms, and soil is essential for the survival and activities of terrestrial organisms. Ecosystems are strongly dependent on, and also affect, the properties of their soils. Under these conditions, ecosystems provide various services to humans (known as ecosystem services). The Millennium Ecosystem Assessment (2005) report describes four main classes of ecosystem services:

1. Provisioning services such as food, water, timber, and fiber.
2. Regulating services that affect climate, floods, disease, wastes, and water quality
3. Cultural services that provide recreational, aesthetic, and spiritual benefits
4. Supporting services such as soil formation, photosynthesis, and nutrient cycling

Among the supporting services is "soil formation," which includes the functions of biota and organic and inorganic constituents in the soil. Digestion of biogenic residues and reuse of nutrient elements constitute "photosynthesis and nutrient cycling." Soil supports the digestion of biogenic residues and the reuse of nutrient elements, and it plays a nutrient-retention role, at least temporarily. Humans utilize the supporting services to obtain provisioning-service products, such as foods,

M. Nanzyo, H. Kanno, *Inorganic Constituents in Soil*,
https://doi.org/10.1007/978-981-13-1214-4_1

fibers, and woods, although agricultural production is often enhanced by the application of fertilizer.

Related to ecosystem services, soils are also involved in landscapes. A landscape is formed by the interactions among soil, topography, climate, and communities of organisms (plants and animals). The ecosystem services provided by landscapes may be recognized as cultural services, for example, areas for recreation and motifs in paintings. Different landscapes are often underlain by different soils. The regulating services offered by soils include the holding and movement of water, maintenance of water quality, and adsorption and desorption of solutes.

Soil-supported ecosystem services contribute greatly to our survival and to environmental sustainability. To maintain and improve our physical and cultural health, we must understand the roles of soil in ecosystem services and develop methods that conserve and facilitate soil functions.

1.2 Elements Important for Ecosystem Services and Environmental Factors Affecting the Behavior of Inorganic Constituents in Soil

As the primary producers in ecosystems, plants absorb inorganic nutrients from soil. Plants and animals occupy the natural, agricultural, and urban landscapes formed by soils. Hence, the behavior and cycling of nutrient elements in soils and ecosystems is critically important. Nutrient elements (elements that are essential or beneficial to plants and/or animals) are transferred to animals from plants through the food chain. Therefore, feed production must meet the nutrient-element demands of animals. Cycling of these nutrient elements is affected by environmental factors such as soil properties, temperature, water, redox condition, carbon dioxide, and light.

1.2.1 Elements

Concentrations of more than 50 soil elements can now be determined by atomic absorption photometry, X-ray fluorescence spectrometry, inductively coupled plasma mass spectrometry, and other techniques (Yamasaki 1996, Takeda et al. 2004). Concentrations of the major soil elements, Si, Al, Fe, Ca, Mg, Na, K, Mn, Ti, and P, are frequently measured. Other major elements, O, H, C, N, S, F, and Cl, exist in soils as oxides, hydroxides, carbonates, nitrates, sulfates, sulfides, fluorides, chlorides, and other compound forms depending on the soil properties. The major elements in soil organic matter are C, N, S, O, and H.

The major essential elements for plants are C, H, O, N, P, K, Ca, Mg, and S. Plants also require trace amounts of Fe, Mn, Cu, Zn, B, Mo, and Cl (Fig. 1.1) (Marshner 1995). Ni is another essential trace element for plants (Asher 1991), and

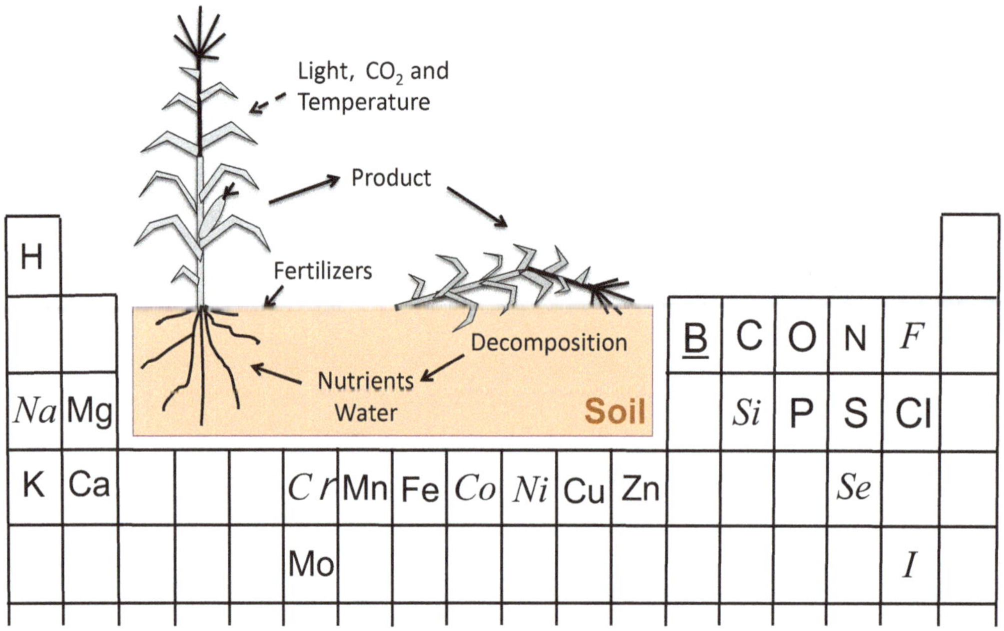

Fig. 1.1 Essential nutrient elements for plants and animals (underlined and italicized elements are essential only to plants or animals, respectively; all others are essential to both plants and animals). Animals receive some of their essential elements through plants, which absorb them from soil

Si and Na are beneficial to some plants. In agricultural plant production, it is essential to provide these elements to plants in appropriate amounts from the viewpoints of both economics and environmental conservation. Excessive application, especially of N and P, must be avoided because these elements cause eutrophication problems when released into rivers, lakes, or bays.

Essential elements for animals include F, Cr, Co, Se, and I (Nielsen 1984; Haenleina and Ankeb 2011). For example, animals in New Zealand have been reported as Co deficient in the case that their animal feeds were produced in Co-deficient soil (Lee 1974).

When present in appropriate abundances, the essential and beneficial elements in soil confer positive effects on ecosystem services. In contrast, excessive concentrations of Cd, Cu, As, Hg, Zn, Cr, and other heavy metals in soils are detrimental to ecosystem health. Excessive concentrations of labile Al (Adams 1984) and Ni have negative effects on sensitive plants.

1.2.2 Environmental Factors

The essential, beneficial, and other elements in soil exist as organic or inorganic constituents with various stabilities. The chemical forms, and quantities of these soil constituents depend on environmental factors such as temperature, moisture, activities of organisms, topography, time, and redox conditions (which are themselves dependent on moisture, topography, drainage, and activities of organisms). The environmental factors also include the properties of soil constituents, because each soil constituent interacts with other soil constituents and the reactivity of soil constituents is highly variable. The abundance of some soil constituents (such as allophane and imogolite) is related to the parent materials and the weathering conditions. The formation of allophane and imogolite from volcanic ash is favored by moist climate and good drainage conditions and impeded by dry climate and submerged conditions (as also described in Chap. 4). The first half of this monograph concisely describes the major inorganic soil constituents with visual aids. The latter chapters describe the behaviors of inorganic constituents under various soil environmental conditions.

Chapter 2 focuses mainly on the primary minerals in soil, that is, minerals that have not been chemically altered since their deposition and crystallization from molten lava. Primary minerals are abundant in the silt-and-sand fraction (particle-size fraction with diameter > 2 μm) of soil. The primary minerals provide physical strength to soil and contribute to soil formation through dissolution or weathering at various speeds. The elements released by the weathering of primary minerals include nutrients for organisms.

Chapter 3 introduces the secondary minerals, which are more active in soil than the primary minerals. Exchangeable cations are a subset of secondary minerals with changeable composition and soil-dependent characteristics. For example, acidic soils contain exchangeable Al, whereas some alkaline soils include exchangeable Na. The composition of exchangeable cations is easily affected by solute changes in soil water, although it is not easily affected by simple dilution using pure water. The composition of exchangeable cations also affects the physical properties of soil (Baver 1928). The retention of nutrient cations contributes to plant production. Types of these minerals can be identified by changes in the basal spacing with exchangeable cations and accommodation of organic molecules at the interlayer site.

Chapter 4 discusses the non-crystalline soil constituents that characterize many areas of volcanic activity. Under good drainage conditions, volcanic glasses alter to non-crystalline materials such as allophane, imogolite, and Al–humus complexes. All of these materials are highly reactive with phosphate, show variable charge properties, and have characteristic physical properties. Phytoliths are another non-crystalline silica material frequently found in humus-rich horizon soils. Phytoliths are a possible source of Si for plants.

Reducing conditions result from submergence of soil and microbial activity. The chemical forms of redox-sensitive elements, including many nutrient elements, differ under reducing and oxidizing conditions. Redox-sensitive inorganic soil constituents are discussed in Chap. 5.

Chapter 6 introduces three topics related to inorganic constituents in soil, namely, tsunami-affected soils, Cs-affected soils, and phosphate reactions in the soil–plant system. The huge tsunami that struck the Pacific coast of eastern Japan in March of 2011 inundated coastal areas with a large volume of seawater. Tsunamis affect soil mainly by erosion, deposition, and by increasing the salt concentration in soil water and the exchangeable Na. Although exchangeable cations are a part of silicate minerals, they are highly sensitive to solute changes in soil water. The 2011 tsunami also destroyed the Fukushima Daiichi Nuclear Power Plant, depositing large amounts of radioactive elements on the soil surface. The behavior of radiocesium in the soil–water system is described. Finally, this chapter discusses the phosphates related to soil–plant systems. The behavior of phosphates, whether inherited from parent materials or applied as fertilizers, is closely related to the inorganic constituents of the soil.

1.3 Purpose

This monograph serves two purposes. Accompanied by visual aids, it first introduces the simplified fundamentals of inorganic soil constituents for students of soil science and interested researchers in other disciplines. Scientific information pertaining to inorganic soil constituents has become increasingly important as concern for the environment has increased. The second purpose of the monograph is to update topics on non-crystalline inorganic soil constituents, the effects of redox reactions, and the effects of disasters on the inorganic constituents of soil. These topics appear in the latter chapters.

Many complete texts and references about minerals in soil are already available (Dixon and Weed 1989; Dixon and Schulze 2002; Huang et al. 2012; Deer et al. 2013). These materials systematically describe the crystallography, properties, formation, and occurrence of the mineral constituents and are recommended for further study.

1.4 Methods

This section, except for the final paragraph, describes the various methods mainly used to collect the results presented in succeeding chapters. Particle size fractionation is effective for studying the inorganic constituents in soil. Primary and secondary minerals were prepared by routine treatments of soil samples, such as air-drying, gentle grinding with a mortar, dry sieving, H_2O_2 digestion, and ultrasonic treatment (Gee and Bauder 1986). After dispersion, the particle sizes were fractionated by wet sieving and siphoning. Dispersion was maximized by adjusting the pH. Alkaline conditions (pH = 10.5) are effective for crystalline clays. Primary minerals were treated with dithionite-citrate-bicarbonate (DCB) when necessary. DCB treatment was also used for X-ray diffraction analysis of the clay fraction (Harris and White 2008). Heavy and light minerals were separated using a heavy liquid. This method is effective, but it is affected by composite mineral particles, which are not rare in soils.

The landscapes and soil profiles at the sampling sites of inorganic constituents were photographed, microphotographs of the inorganic constituents in soil were taken through a Leica M 205C stereomicroscope, and the isotropic or anisotropic properties (Lynn et al. 2008) of the constituents were analyzed in optical micrographs taken under a Nikon ECLIPSE E600 POL microscopic system.

The micro-scale morphological properties of soil inorganic constituents were revealed in scanning electron microscope (SEM) (White 2008) and transmission electron microscope (TEM) (Elsass et al. 2008) images. For SEM observation, air-dried sample particles were held on double-sided sticky tape. Polished sections were used to reveal the two-dimensional structures of inorganic soil particles, soil clods, and rice–root bundles. To avoid charging, the samples were coated with vacuum-evaporated carbon. Alternatively, Pt–Pd coating was applied if the carbon coating was insufficient. For TEM observation, sample particles were supported on a copper mesh by a collodion membrane. Allophane and imogolite were observed on a TEM microgrid. TEM photographs were obtained by a Quemesa digital camera. Both the SEM and TEM observations were carried out under high vacuum. For morphological observation and energy dispersive X-ray (EDX) analyses (Guillemette 2008), the accelerating voltage of the SEM was typically set to 15 kV, but it was sometimes set to 3 kV for detailed morphological observations. Most of the SEM and TEM observations were performed on Hitachi SU8000 and H-7650 instruments, respectively.

The elemental compositions of inorganic constituent particles and the distribution of elements can effectively identify the particles and may reveal chemical reaction products. This monograph presents many EDX spectra as visual examples of the element abundances of inorganic soil constituents. Elemental compositions were analyzed from the characteristic X-ray spectra of the samples (Fig. 1.2) obtained by EDX. Metals heavier than Fe are rare among the inorganic constituents of most soils. The relative abundances of the elements can be roughly estimated from the heights of the peaks of the characteristic X-rays in the spectra although the elemental composition is accurately calculated in quantitative analysis using an equipped software. In early work, EDX analyses were usually performed on a Kevex apparatus, but more recent analyses were performed using an EDAX Apollo XV.

Points of attention for EDX analytical results include overlapping of characteristic X-rays, the position of the X-ray detector, and some others. Among the overlapping of characteristic X-rays mentioned in the EDX manual, $FK_\alpha - FeL_\alpha$ and $PK_\alpha - ZrL_\alpha$ may sometimes be encountered when observing inorganic constituents in soil. As the X-ray detector is installed diagonally upward from the sample, unevenness in the sample surface affects the effectiveness of the detector for detecting characteristic X-rays, and the X-ray intensity from the opposite side of the sample is weak. Also, shadows are formed in the element maps of particle samples of soil inorganic constituents. To avoid the effects of uneven samples, it is necessary to use adequately flat or polished sections with soil inorganic constituents embedded in resin.

X-ray diffraction patterns of powder samples were acquired by a Rigaku MiniFlex X-ray diffractometer using the CuK_α (30 kV, 15 mA) line at a scanning speed of 2° per minute. For this purpose, oriented samples were prepared on glass slides using Mg^{2+}- and K^+-saturated samples at room temperature, and their X-ray

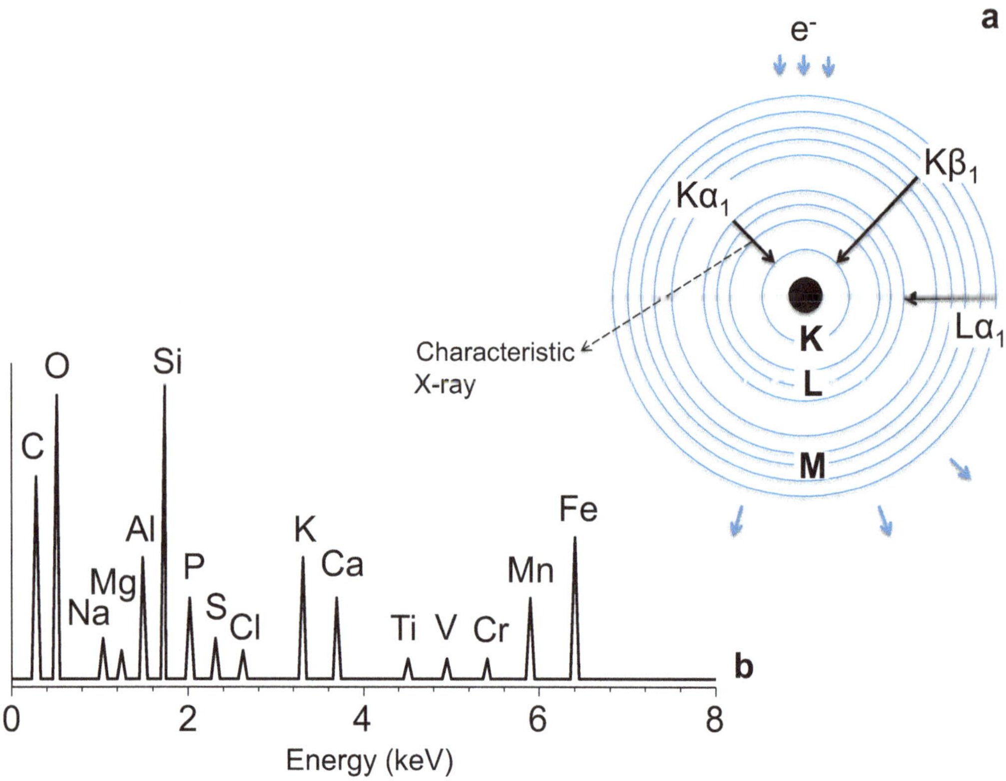

Fig. 1.2 Energy dispersive X-ray spectrum. (**a**) Schematic of an atom struck by an electron beam, (**b**) a model EDX spectrum in which each element peak (K_α) is at its specific energy but the peak height is arbitrary

diffraction patterns were obtained. Changes in basal spacing of layer silicates were recorded after solvation of the Mg^{2+}-saturated samples with glycerin and heating of the K^+-saturated samples at 300 and 550 °C for 1 h. The clay mineral composition was evaluated from the basal-spacing changes of the layered silicates (Harris and White 2008).

The element concentrations in digested or extracted solutions were determined by colorimetric spectroscopy, atomic absorption spectroscopy, and inductively coupled plasma spectroscopy. Radiocesium concentrations in soil were determined by gamma-ray spectrometry (ORTEC, GEM, and DSPEC jr 2.0, 2000s). The radioactivity of soil mineral particles and cross sections of muddy tsunami deposits embedded in resin were detected by an imaging plate (BAS-5000, FUJIFILM Co., Ltd.).

Synchrotron radiation reveals the chemical forms of elements by X-ray absorption near edge structure (XANES) and extended X-ray absorption fine structure (EXAFS) (Kelly et al. 2008). Originally employed for heavy-element analysis, XANES and EXAFS are increasingly being applied to lighter elements. High-resolution X-ray ptychography (accurate to 10 nm), X-ray computed tomography, XANES and EXAFS are expected to become the methods of choice in future analyses of lighter elements (Ajiboye et al. 2008; Trinh et al. 2017) in soil-plant systems.

References

Adams F (ed) (1984) Soil acidity and liming, 2nd edn. Agronomy; no.12, ASA-CSSA-SSSA, Madison

Ajiboye B, Akinremi OO, Hu Y, Jurgensen A (2008) XANES speciation of phosphorus in organically amended and fertilized vertisol and mollisol. Soil Sci Soc Am J 72:1256–1262

Asher CJ (1991) Beneficial elements, functional nutrients, and possible new essential elements. In: Mortvedt JJ, Cox FR, Shuman LM, Welch RM (eds) Micronutrients in agriculture, SSSA book series, no. 4, 2nd edn. SSSA, Madison, pp 703–723

Baver LD (1928) The relation of exchangeable cations to the physical properties of soils. Agron J 20:921–941

Deer WA, Howie RA, Zussman J (2013) An introduction to the rock-forming minerals, 3rd edn. Mineralogical Society, London

Dixon JB, Schulze DG (eds) (2002) Soil mineralogy with environmental applications. SSSA book series, no. 7. SSSA, Madison

Dixon JB, Weed SB (eds) (1989) Minerals in soil environment. SSSA book series no. 1. SSSA, Madison

Elsass F, Chenu C, Tessier D (2008) Transmission electron microscopy for soil samples: preparation methods and use. In: Ulery A, Vepraskas M, Wilding L (eds) Methods of soil analysis. Part 5. Mineralogical methods. SSSA book series, no. 5. SSSA, Madison, pp 235–268

Gee GW, Bauder JW (1986) Particle-size analysis. In: Klute A (ed) Methods of soil analysis: part 1 – physical and mineralogical methods, Agronomy monograph no. 9, 2nd edn. SSSA, Madison, pp 383–411

Guillemette RN (2008) Electron microprobe techniques. In: Ulery A, Vepraskas M, Wilding L (eds) Methods of soil analysis. Part 5. Mineralogical methods. SSSA book series, no. 5. SSSA, Madison, pp 335–365

Haenleina GFW, Ankeb M (2011) Mineral and trace element research in goats: a review. Small Rumin Res 95:2–19

Harris W, White GN (2008) X-ray diffraction techniques for soil mineral identification. In: Ulery A, Vepraskas M, Wilding L (eds) Methods of soil analysis. Part 5. Mineralogical methods. SSSA book series, no. 5. SSSA, Madison, pp 81–115

Huang PM, Li Y, Sumner ME (eds) (2012) Handbook of soil sciences: properties and processes, 2nd edn. CRC Press/Taylor & Francis Group, Boca Raton/London/New York

Kelly SD, Hesterberg D, Ravel B (2008) Analysis of soil and minerals using X-ray absorption spectroscopy. In: Methods of soil analysis part 5 – mineralogical methods, SSSA book series no.5. SSSA, Madison, pp 387–463

Lee HJ (1974) Trace elements in animal production. In: Nicholas DJD, Egan AR (eds) Trace elements in soil-plant-animals systems. Academic, New York/San Fransisco/London, pp 39–54

Lynn W, Thomas JE, Moody LE (2008) Petrographic microscope techniques for identifying soil minerals in grain mounts. In: Ulery A, Vepraskas M, Wilding L (eds) Methods of soil analysis. Part 5. Mineralogical methods. SSSA book series, no. 5. SSSA, Madison, pp 161–190

Marschner H (1995) Mineral nutrition of higher plants. Academic, London

Millenium Ecosystem Assessment (2005) Ecosystems and human well-being: synthesis. 137p. Island Press, Washington, DC

Nielsen FH (1984) Ultratrace elements in nutrition. Annu Rev Nutr 4:21–41

Takeda A, Kimura K, Yamasaki S (2004) Analysis of 57 elements in Japanese soils, with special reference to soil group and agricultural use. Geoderma 119:291–307

Trinh TK, Nguyen TTH, Nguyen TN, Wu TY, Meharg AA, Nguyen MN (2017) Characterization and dissolution properties of phytolith occluded phosphorus in rice straw. Soil Tillage Res 171:19–24

White GN (2008) Scanning electron microscopy. In: Ulery A, Vepraskas M, Wilding L (eds) Methods of soil analysis. Part 5. Mineralogical methods. SSSA book series, no. 5. SSSA, Madison, pp 269–297

Yamasaki S-I (1996) Inductively coupled plasma mass spectrometry. In: Boutton TW, Yamasaki S-I (eds) Mass spectrometry of soils. Marcel Dekker, Inc., New York/Basel/Hong Kong, pp 459–491

Chapter 2
Primary Minerals

Abstract This chapter introduces the primary minerals that are relatively common in soils. It first presents the accepted views on the elemental compositions of the Earth's crust, rocks, and minerals. Soils at the top of the Earth's crust are also within the rock cycle. Silicate and silica minerals, which constitute more than 90% of the minerals in the Earth' crust, are outlined. Samples of relatively un-weathered and weathered primary minerals were obtained from new volcanic ash and soils derived from granitic rocks, respectively. Quartz is highly resistant to weathering, whereas biotite in soil is altered in moist climates. The composition of primary minerals in soils is affected by the types of parent rocks, weathering, sorting, and other soil-forming factors, resulting in mineral compositions that deviate from the average mineral composition of the Earth's crust.

2.1 Introduction

Ranging from clay to rock fragment, soil particles have a wide size distribution. Minerals in soils are divided conceptually into primary and secondary minerals. According to the Glossary of Soil Science Terms (Glossary of soil science terms committee 2008), a primary mineral is a mineral that has not been altered chemically since its crystallization from molten lava and deposition. A mineral is defined as an inorganically formed, naturally occurring homogeneous solid with a definite chemical composition and an ordered atomic arrangement. These definitions are followed in this monograph.

Soils form by widely different processes, and their states range from un-weathered to highly weathered. They thus show various compositions of primary and secondary materials. The particle-size fraction of most primary minerals is the larger than 2 μm fraction, which includes silt, sand, and gravel. Primary minerals can be separated from soil by the particle-size fractionation method described in the Sect. 1.4.

The major primary minerals in soil are silicate and silica minerals. Other minerals include titanomagnetite, other iron minerals, and apatite. The sand fraction of soils

M. Nanzyo, H. Kanno, *Inorganic Constituents in Soil*,
https://doi.org/10.1007/978-981-13-1214-4_2

includes non-crystalline inorganic constituents, such as volcanic glasses. Volcanic glasses and apatite are introduced in Chap. 4 and Sect. 6.3, respectively.

Particles larger than silt includes fine rock fragments, complex particles of different minerals, and partially weathered minerals. After introducing the major primary minerals in soils, this chapter exemplifies the mineral composition of fine rock fragments on a polished section, and partially weathered minerals.

2.2 Average Mineral Composition of the Earth's Crust

Earth's surface is naturally mobile. The mobile layer is made up of plates comprising the Earth's crust and uppermost mantle. On average, the Earth's crust is 40 km thick on continents and 6 km thick in the oceans. The upper mantle consists mainly of peridotite, whereas the crust consists of igneous rocks, sedimentary rocks, and metamorphic rocks (Fig. 2.1). These three rock families of the Earth's crust can interchange through diastrophism (the rock cycle, Fig. 2.2) (Skinner and Mruck 2011). The phases of the rock cycle can undertake various shortcuts. The rock cycle is active in the subduction zone of a plate and in the collision zones of continents, but it is relatively inactive in stable continental crust. However, on the surface of the continental crust, soil formation processes are continuously active.

According to the estimated average element composition of the Earth's crust, oxygen is the most abundant element, followed by Si and Al (see Fig. 2.1). The crust consists of 65% igneous rocks, 27% metamorphic rocks, and 8% sedimentary rocks (Fig. 2.1). Approximately two-thirds of the igneous rocks are basic rocks; neutral and acidic rocks constitute approximately one-sixth each. The very surface of the Earth's crust is dominated by sedimentary rocks, which are strongly affected by weathering, erosion, transportation, and deposition (Fig. 2.2).

The rocks of the Earth's crust are dominated by plagioclase, followed by quartz, alkali feldspar, and other silicates. Collectively, these minerals constitute approximately 92% of the rock material (Fig. 2.1). Other minerals are non-silicate minerals such as carbonates, sulfates, phosphates, sulfides, fluorides, and chlorides. The alteration of rocks and minerals depends on the soil formation factors, which vary across the surface of the Earth.

2.3 Silicate and Silica Minerals

2.3.1 Grouping of Silicate and Silica Minerals

Silicate and silica minerals are grouped into six types based on the bonding structure of their silicate tetrahedrons. Examples and the ideal chemical formulas of the six types of silicates are summarized in Table 2.1. The model structures of five silicate types are shown in Fig. 2.3. Nesosilicates are characterized by single SiO_4 tetrahedra

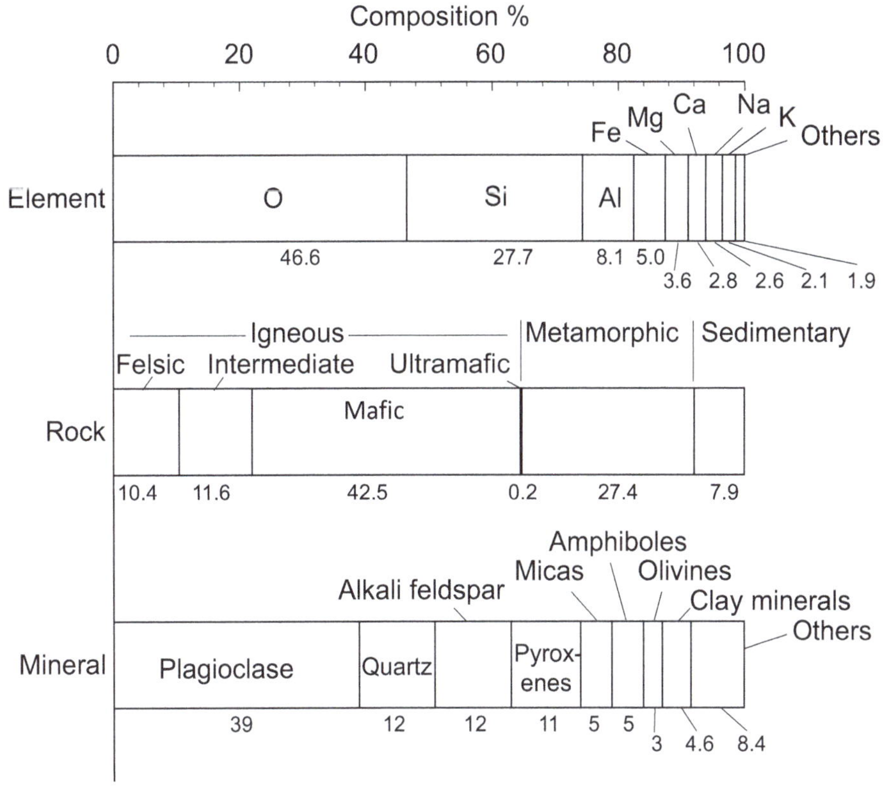

Fig. 2.1 Average element and rock and mineral (Wyllie 1971) compositions of the Earth's crust

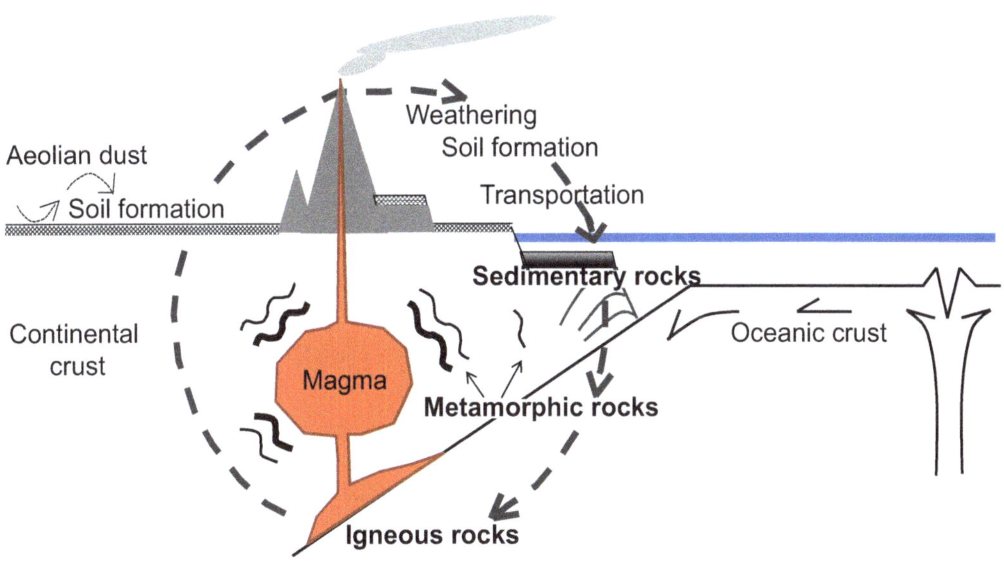

Fig. 2.2 Schematic of the rock cycle. Various shortcuts are possible in this cycle

with no Si–O–Si bonding. Other silicates are constructed from units of two or more connected SiO_4 tetrahedra. Sorosilicates are dimers of SiO_4 tetrahedra. Cyclosilicates have a ring structure composed of 3–6 SiO_4 tetrahedra. Inosilicates

Table 2.1 Grouping of silicate and silica minerals

	Subclass name	Characteristics		Example	Ideal formula
1	Nesosilicates	Lone tetrahedron	$[SiO_4]^{4-}$	Olivine	$(Mg, Fe^{2+})_2SiO_4$
2	Sorosilicates	Double tetrahedra	$[Si_2O_7]^{6-}$	Vesuvianite	$Ca_{19}(Al, Fe)_{10}(Mg, Fe)_3(Si_2O_7)_4(SiO_4)_{10}(O, OH, F)_{10}$
3	Cyclosilicates	Ring silicates	$[Si_nO_{3n}]^{2n-}$	Tourmaline	$NaMg_3Al_6(Si_6O_{18})(BO_3)_3(OH)_3(OH, F)$ (dravite)
4	Inosilicates	Single chain silicates	$[Si_nO_{3n}]^{2n-}$	Pyroxene	$(Ca, Mg, Fe^{2+}, Al)_2(Si, Al)_2O_6$ (augite)
		Double chain silicates	$[Si_{4n}O_{11n}]^{6n-}$	Amphibole	$Ca_2(Mg, Fe^{2+})_4 Al [Si_7Al]O_{22}(OH)_2$ (magnesiohornblende)
5	Phyllosilicates	Sheet silicates	$[Si_{2n}O_{5n}]^{2n-}$	Muscovite	$K_2Al_4(Si_6Al_2)O_{20}(OH)_4$
				Biotite	$K_2(Mg, Fe)_6(Si_6Al_2)O_{20}(OH)_4$
				Clay minerals	See Chap. 3
6	Tectosilicates	3D framework	$[Al_xSi_yO_{2(x+y)}]^{x-}$	Quartz	SiO_2
				Feldspar	$NaAlSi_3O_8$ (albite)

consist of linear chains of SiO_4 tetrahedra. The chains may be single or double. In phyllosilicates, the SiO_4 tetrahedra are assembled into 6-membered rings, which connect and spread two-dimensionally into a sheet-like structure. Tectosilicates are three-dimensional assemblages of SiO_4 tetrahedra. In soils, nesosilicates, inosilicates, phyllosilicates, and tectosilicates are common to abundant and present as primary and clay minerals. Allophane and imogolite are also grouped in the nesosilicates, although these two are non-crystalline.

2.3.2 *Examples of Silicate and Silica Minerals in Soil*

This section presents examples of the silicate and silica minerals frequently found in soils. Several examples of un-weathered mineral particles were taken from new volcanic ash deposits. Partially weathered minerals are so common in soils that a few examples of them are also included in this section.

2.3.2.1 Silicate Minerals

Silicate minerals are various salts of silicate anions (Fig. 2.3). The cations are Al, Mg, Fe, Ti, Na, K, Ca, and other elements. In the following discussion, the silicate minerals are introduced in order of increasing complexity of their silicate framework.

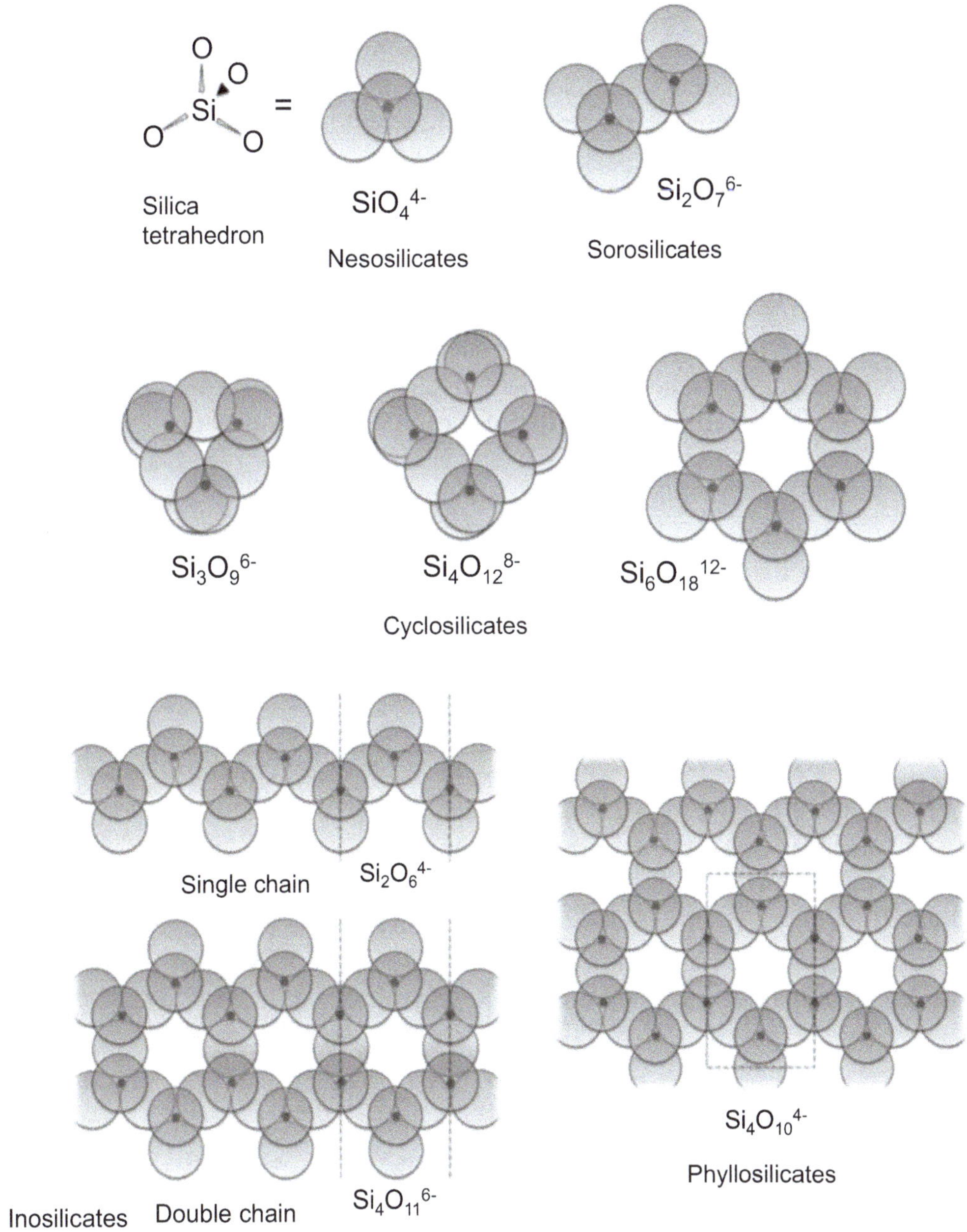

Fig. 2.3 Model structures of silicate and silica minerals. For tectosilicates, refer to Deer et al. (2013)

The minerals are characterized by their EDX spectra and the X-ray diffraction (XRD) patterns of their powder samples. These data are presented along with an optical micrograph and an SEM image of each mineral. The SEM images show the detailed morphological properties of the material. Reference EDX spectra-mimic graphs showing the reference elemental compositions (atomic number ratios) of each mineral are also provided. The horizontal axes of the EDX spectra-mimic graphs are matched with those of the EDX spectra so that readers can easily compare the exemplified mineral with the reference data.

Olivine

Olivine is a light-yellow to yellow-brown or olive-green nesosilicate (Fig. 2.4a). It is sometimes found in basic rocks or scoria. For every Si atom, there are two atoms of Mg or Fe and four atoms of oxygen. The highest and lowest Mg end-members are Mg_2SiO_4 and $Fe(II)_2SiO_4$, respectively. Panels (d), (e), and (f) of Fig. 2.4 show the chemical compositions of the olivine members with the maximum, intermediate, and minimum number of Mg atoms, respectively, from among the 115 analytical data for the Earth presented in Deer et al. (1997a), who reviewed the mineralogical properties of olivine. Huang (1989) detailed the structural properties, natural occurrences, equilibrium environment, and physicochemical properties of olivine in soil.

The olivine particles shown in panels (a) and (b) of Fig. 2.4 were picked by tweezers from the 0.5–0.2 mm fraction of the 2A1 horizon of the pedon shown in Fig. 4.5a. The EDX spectrum in Fig. 2.4c is similar to that of Fig. 2.4e, suggesting that this example has a composition close to the intermediate chemical composition of olivine. The powder XRD pattern (Fig. 2.4g) approximates the reference pattern (Fig. 2.4h, from Brindley and Brown 1980).

Fig. 2.4 Olivine particles from a scoria deposit from Mt. Fuji. (**a**) Optical photograph, (**b**) SEM image, (**c**) EDX spectrum. (**d**, **e**, and **f**) the elemental compositions of the olivine group minerals having the maximum, intermediate, and minimum number of Mg atoms, respectively, shown to mimic EDX spectra, (**g**) the powder XRD pattern of the olivine particles, (**h**) the reference powder XRD pattern (Brindley and Brown 1980). These olivine samples were separated from the soil profile shown in Fig. 4.5a

Pyroxene

Pyroxenes are single chain silicates that are linked laterally by cations such as Mg, Fe, Ca, and others. Pyroxenes are grouped by their chemical compositions into Mg–Fe pyroxene, Ca pyroxene, Na pyroxene, and others. Pyroxenes are also grouped into orthopyroxenes and clinopyroxenes according to their crystal system. As major pyroxenes, Mg–Fe pyroxene (enstatite–ferrosilite) of the orthopyroxenes and augite of the clinopyroxenes are introduced. According to its grouping by chemical composition, augite is one of the Ca pyroxenes. Deer et al. (1997b, 2013) described details of the pyroxenes.

Panels (f), (g), and (h) of Fig. 2.5 show the chemical compositions of the orthopyroxene members having the maximum, intermediate, and minimum number of Mg atoms, respectively, from among the 73 examples listed by Deer et al. (1997b). The vertical axis shows the number of cations per 6 oxygens. The maximum Mg member contains a small number of Fe atoms. The minimum Mg member contains a large number of Fe atoms and a small number of Ca atoms. The example shown in Fig. 2.5c corresponds to an orthopyroxene member between that of Fig. 2.5f and that of Fig. 2.5g. The powder X-ray diffraction pattern of the orthopyroxene sample is close to the reference, 31–634 (Hypersthene) of Joint committee on powder diffraction standards (1986).

The orthopyroxene shown in Fig. 2.5c was separated from the 2–0.25 mm fraction of the Tarumae-a (Ta-a) tephra sampled at Oiwake (Iburi Subprefecture, Hokkaido, Japan) (Mizuno et al. 2008) near the pedon site shown in Fig. 2.5a. The Ta-a tephra, erupted in 1739 from Mt. Tarumae, corresponds to the C horizon labeled as H5–2 of the soil profile shown in Fig. 2.5a. The soil color of the H5–2 horizon is whitish and the soil texture is sand, indicating that weathering is weak. Figure 2.5b shows not only light brown orthopyroxene but also light green augite, beige pumice-like volcanic glass, whitish feldspar, and some other grains from the tephra. Many of the crystalline minerals are partly or almost wholly covered with colorless volcanic glass.

Figure 2.6a shows augite, one of the clinopyroxenes, separated from the same fraction of the Ta-a tephra as the above-mentioned orthopyroxene. The color of the augite is light-green, and the grains are partly covered with colorless volcanic glass. The EDX spectra-mimic graphs Fig. 2.6d–f show the number of cations per 6 oxygen atoms of the augite members having the maximum, intermediate, and minimum number of Mg atoms, respectively, from among the 101 samples shown by Deer et al. (1997b). With a decrease in the Mg concentration, the concentration of Fe tends to increase. The number of Ca ions does not change very much compared with those of Mg and Fe among these members. The EDX spectrum (Fig. 2.6c) is close to Fig. 2.6e.

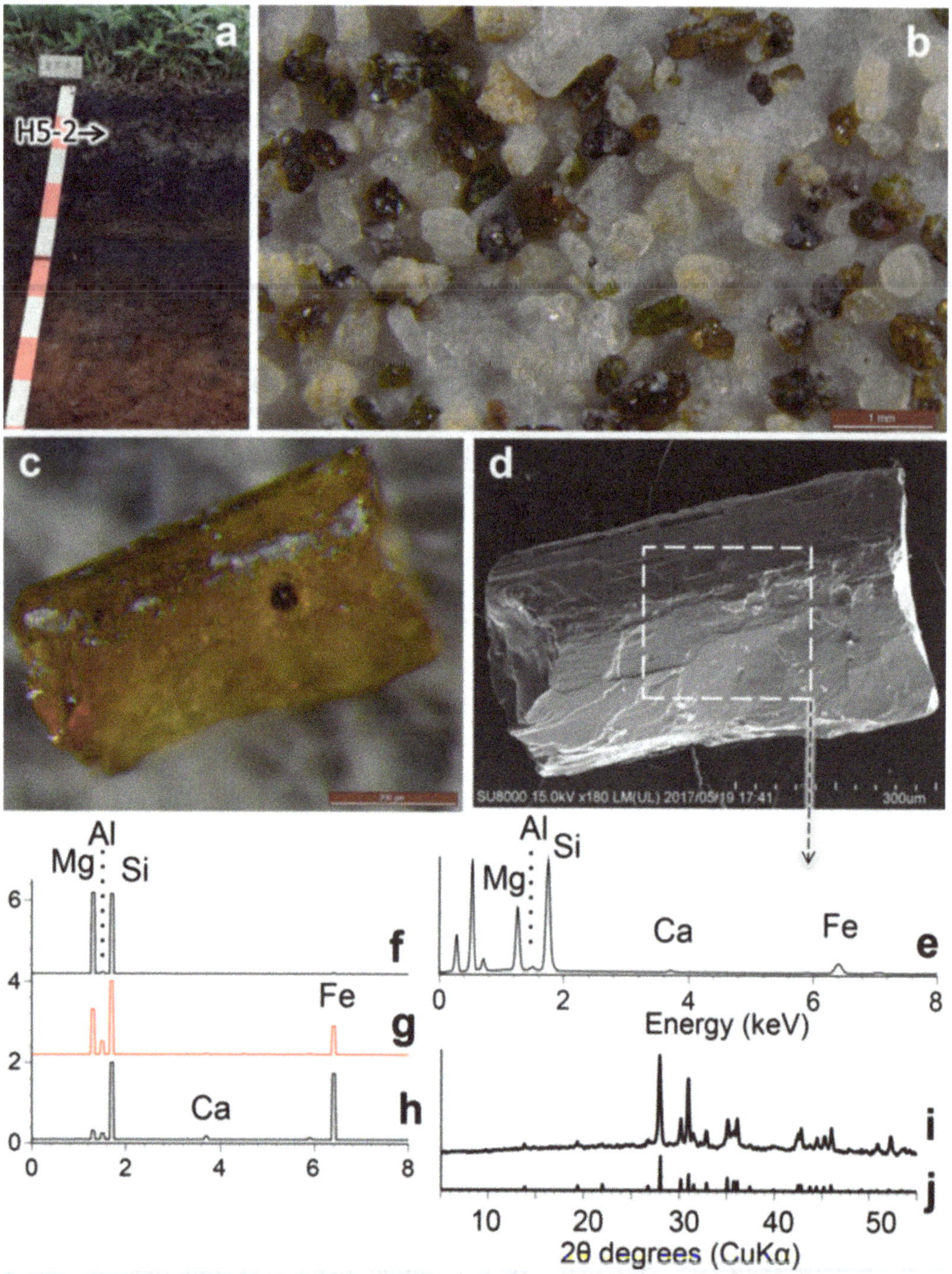

Fig. 2.5 Orthopyroxene of the Tarumae-a (Ta-a) tephra. (**a**) Ta-a tephra labeled as H5–2 in the soil profile, (**b**) optical micrograph of the 2–0.25 mm fraction separated from the Ta-a tephra, (**c** and **d**) optical micrograph and SEM image of an orthopyroxene particle, respectively, (**e**) EDX spectrum of the dashed area of (**d**), (**f**, **g**, and **h**) EDX spectra-mimic graphs showing the elemental compositions of orthopyroxenes with the maximum, intermediate, and minimum number of Mg atoms (Deer et al. 1997b), (**i** and **j**) powder XRD pattern of the orthopyroxene particles and reference powder XRD pattern, respectively (Joint committee on powder diffraction standards 1986)

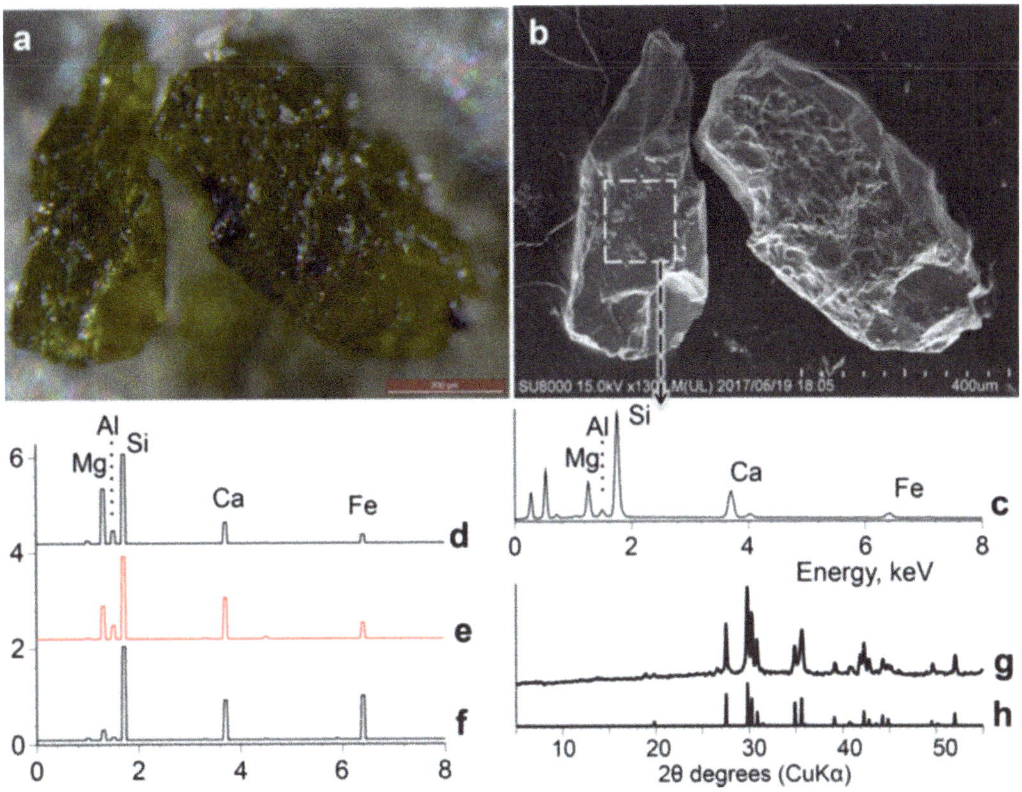

Fig. 2.6 Augite separated from the Ta-a tephra. (**a**) Optical micrograph, (**b**) SEM image of the augite, (**c**): EDX spectrum of the dashed area in (**b**), (**d**, **e**, and **f**) elemental compositions of augites with the maximum, intermediate, and minimum number of Mg ions, respectively (Deer et al. 1997b). (**g**) Powder XRD pattern of the augite particles, (**h**) reference powder XRD pattern (24-203 of Joint committee on powder diffraction standards 1986)

Amphibole

Amphiboles are double chain silicates (Table 2.1). Between the double chains, several cations link the double chain silicates. The general chemical composition is formulated as follows:

$A_{0\text{-}1}B_2C_5T_8O_{22}(OH)_2$

For example, in the Ca-rich amphiboles,
A: Na, K, or vacant
B: Ca, Na, Mn
C: Mg, Fe^{2+}, Fe^{3+}, Al, Ti, Mn, Cr or other
T: Si, Al

The cations at the A, B, C, and T sites play their own roles. The C site cations are sandwiched with two double chain silicates. The sandwiched double chain silicates are laterally linked by the B site cations. The OH group, which is bound to C site cations, can be partially or wholly replaced by F and Cl. In the case of oxyhornblende, the OH group is replaced by oxygen. The T site is the silicate chain.

Depending on the B-site cations, amphiboles are grouped as calcic amphibole, sodic amphibole, sodic-calcic amphibole, and iron-magnesium-manganese amphibole. Hornblendes, commonly found amphiboles, are grouped as calcic amphibole and are members of the magnesio-hornblendes (Mg can be replaced by Fe^{2+} in a wide range) at the lower level of grouping (Deer et al. 1997c).

Hornblendes often occur in volcanic ash. Figure 2.7a, b show dacitic volcanic ash erupted from Mt. Pinatubo (Zambales Mountains, Philippines) in June 1991 (Wolfe and Hoblitt 1996) and re-deposited along large rivers as lahar (mudflow) deposits. In the 0.2–0.5 mm fraction (Fig. 2.7c), hornblendes can be seen as black prismatic particles in a mixture of sponge-like volcanic glass, feldspars, quartz, etc.

The black prismatic particle appears dark-greenish under an optical microscope (Fig. 2.8a). The EDX spectrum (Fig. 2.8c) indicates that Na, Mg, Al, Si, Ca, and Fe are contained in the particle. The EDX spectrum-mimic graphs (Fig. 2.8d–f) show the number of cations per 23 oxygens or 24 (O, OH, F) of the maximum, intermediate, and minimum Mg members from 206 hornblende group minerals listed by Deer et al. (1997c). With decreasing number of Mg atoms, the number of Fe atoms tends to increase. The number of Ca atoms does not vary as much as those of Mg and Fe. The EDX spectrum of Fig. 2.8c corresponds to a member between those of Fig. 2.8d and e. Although the elemental composition of hornblende may be similar to that of augite (Fig. 2.6), the morphological properties, color, and XRD pattern of

Fig. 2.7 Lahar deposit from Mt. Pinatubo, Philippines. (**a**) Landscape of the lahar deposit at the Pasig-Potrero River, (**b**) profile of the lahar deposit, (**c**) an optical micrograph of the 0.2–0.5 mm fraction of the lahar deposit

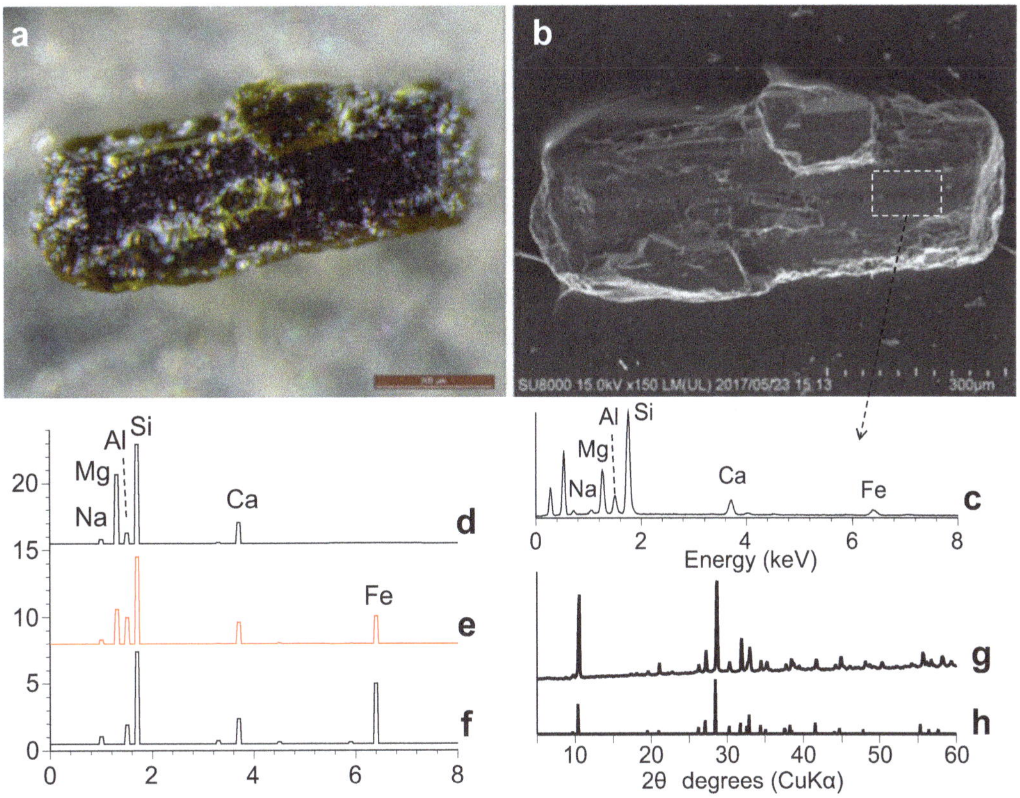

Fig. 2.8 Hornblende particle from the 0.2–0.5 mm fraction of the 1991 Pinatubo tephra taken from the 1992 lahar deposit (Fig. 2.7). (**a**) Optical micrograph, (**b**) SEM image, (**c**) EDX spectrum of the dashed area in (**b**), (**d**, **e**, and **f**) EDX spectrum-mimic graphs showing the elemental compositions of the maximum, intermediate, and minimum Mg members of hornblende, respectively (Deer et al. 1997c), (**g**) powder XRD pattern of the hornblende particles, (**h**) reference powder XRD pattern of hornblende (Joint committee on powder diffraction standards 1986)

hornblende are different from those of augite. The XRD pattern (Fig. 2.8g) resembles 21–149 (Magnesio-hornblende, ferroan) of Joint committee on powder diffraction standards (1986) (Fig. 2.8h).

Micas

Micas are phyllosilicates, which have basal cleavage and flat shape. Micas are one of the major components of igneous, sedimentary, and metamorphic rocks. Abundant members of the micaceous primary minerals are the muscovite and biotite series minerals. The structures of micas are introduced in Chap. 3 with those of other phyllosilicate minerals.

Muscovite

Muscovite (dioctahedral mica) is colorless, transparent, and more resistant to weathering than biotite, a trioctahedral mica. Figure 2.9c shows minerals in the coarse sand fraction separated from a granitic soil in the Republic of Korea. The colorless, transparent, and flat particles are muscovite. As shown in Fig. 2.10a (optical micrograph) and (b) (SEM image), the surface of the muscovite is highly flat, smooth, and appears mostly un-weathered. In contrast, the brown and partly shiny particles appear to be weathered biotite particles in Fig. 2.9c.

Fig. 2.9 Granitic soil in Singun-ri, the Republic of Korea. (**a**) Landscape, (**b**) soil profile and (**c**) 2–0.2 mm fraction of the soil. The 2–0.2 mm fraction of the C horizon (indicated by an arrow in (**b**), 100–125+ cm from the surface) includes feldspar, quartz, white mica, weathered biotite, and other minerals

The aluminum concentration of muscovite is relatively high among the primary minerals. The EDX spectrum-mimic graphs, Fig. 2.10d–f, show the number of cations for the maximum, intermediate, and minimum Al members, respectively, from among 70 muscovites, phengites, and other potassic white micas per 12 (O, OH, F) or 22 anions (Fleet 2003). The number of Al ions in Fig. 2.10d, e is close to that of Si. According to the ideal chemical formula for muscovite (Table 2.1), the number of Al atoms is the same as the number of Si atoms. In phengite, a portion of the Al at the octahedral sites is replaced by Mg^{2+}. The minimum Al member, Fig. 2.10f, is celadonite, in which isomorphous substitution at the octahedral sites is small and the octahedral sites are occupied by Fe^{3+} and Mg^{2+}. The powder XRD pattern, Fig. 2.10g, is close to the reference pattern for muscovite, Fig. 2.10h.

Biotite

The biotite series comprises trioctahedral micas with compositions between, or close to, the phlogopite–annite and eastonite–siderophyllite joins, and also includes tetraferriphlogopite and tetraferriannite because biotite often contains small amounts of Fe^{3+} in its tetrahedral sites (Fleet 2003). Figure 2.11 shows an example of a relatively un-weathered biotite particle from a mudflow deposit of Mt. Pinatubo, Philippines (Fig. 2.7), although biotite particles are not particularly abundant in the mudflow deposit. The shape of the particle is hexagonal and platy. The EDX

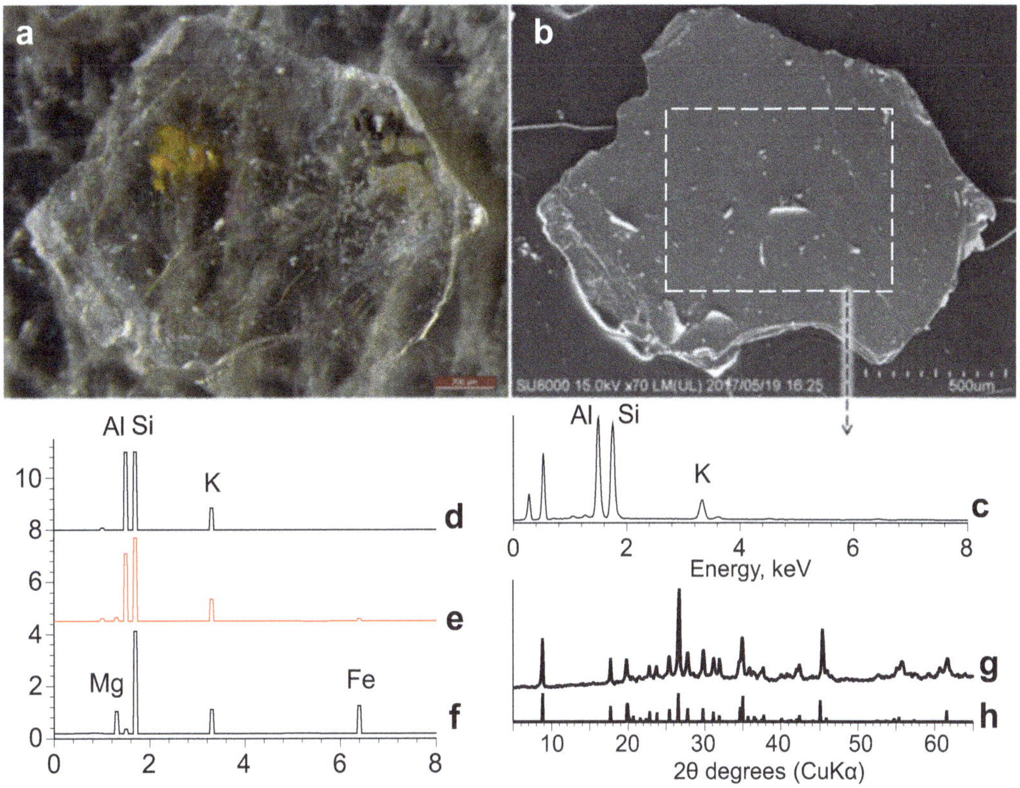

Fig. 2.10 White mica from Singun-ri, the Republic of Korea. (**a**) Optical micrograph, (**b**) SEM image, (**c**) EDX spectrum of the dashed area in (**b**), (**d**, **e**, and **f**) EDX spectrum-mimic graphs showing the elemental compositions of the maximum, intermediate, and minimum Al members of muscovite, respectively, (**g**) powder XRD pattern of the muscovite particles, (**h**) reference powder XRD pattern (Brindley and Brown 1980)

spectrum (Fig. 2.11c) suggests that the chemical composition is between those of the maximum and intermediate Mg members. Although the color is close to black, the chemical composition is not uniform. Partial weathering might have already started because a portion of the K^+ was exchanged with Ca^{2+} during transportation from the original site of deposition by rain and river water.

According to quantitative analyses of several particles similar to the one in Fig. 2.11a, the Mg/(Mg + Fe) atomic ratio ranges between 60 and 70%, indicating that these particles are a Mg-biotite close to phlogopite (Nanzyo et al. 1999). The powder XRD pattern (Fig. 2.11g) is close to that for phlogopite (1 M), Fig. 2.11h (Brindley and Brown 1980).

Further readings for micas are Funning et al. (1989), Thompson and Ukarainczyk (2002).

Feldspars

Feldspars, including plagioclase and alkali feldspar, account for half of the minerals of the Earth's crust (Fig. 2.1). Feldspars are tectosilicates (Table 2.1) and their three end-members are orthoclase, albite, and anorthite (Fig. 2.12). Depending on their formation temperature, feldspars are divided into high- and low-temperature types (Dear et al. 2013). The dotted area of the triangle (Fig. 2.12) shows elemental

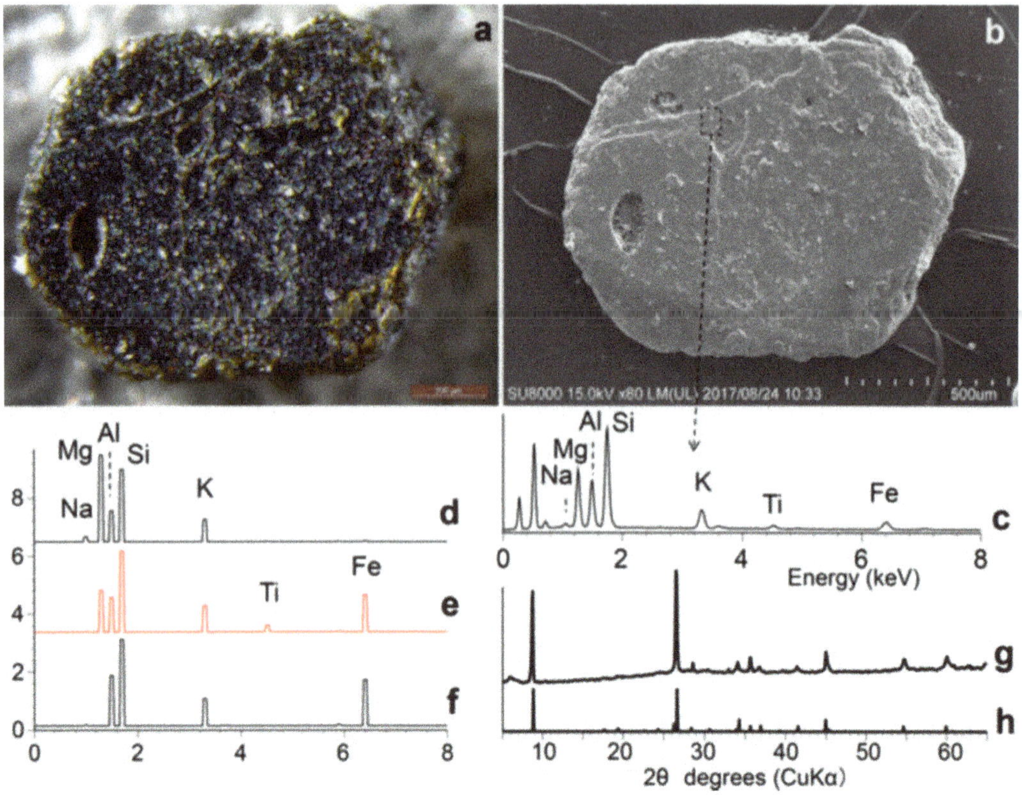

Fig. 2.11 Biotite from a lahar deposit of Mt. Pinatubo. (**a**) Optical micrograph, (**b**) SEM image, (**c**) EDX spectrum of the dashed area in (**b**), (**d**, **e**, and **f**) EDX spectrum-mimic graphs showing the elemental compositions of the maximum, intermediate, and minimum Mg members of the biotite series (112 samples) listed by Fleet (2003), (**g**) powder XRD pattern of the biotite particles, (**h**) reference powder XRD pattern (Brindley and Brown 1980)

Fig. 2.12 Simplified triangle diagram showing the K, Na, and Ca compositions of natural feldspars and their end-members (orthoclase, albite, and anorthite)

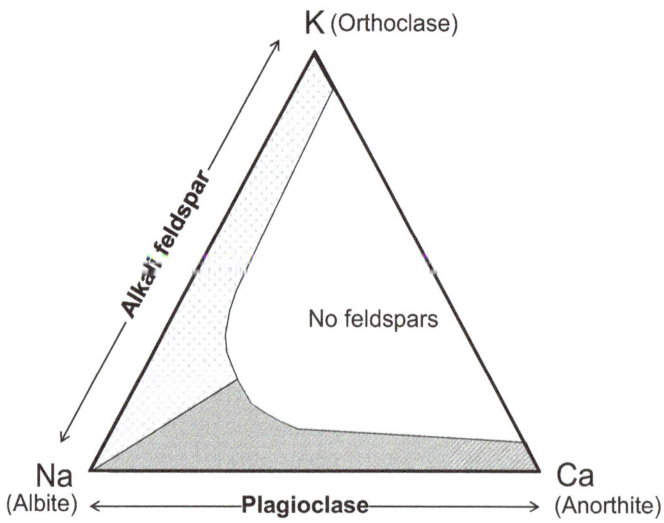

compositions of the alkali feldspars. With slow cooling, the alkali feldspars tend to convert to the two areas near orthoclase and albite. Plagioclase (shaded area of Fig. 2.12) is regarded as a solid solution of two end members, albite and anorthite, and is divided into four general categories. These are oligoclase (Albite: 90–70%), andesine (Albite: 70–50%), labradorite (Albite: 50–30%), and bytownite (Albite: 30–10%). No feldspar exists in the white area. When high-temperature feldspars are cooled rapidly, they may retain the high-temperature phase at ambient temperature. The low-temperature type occurs in the plutonic rock that cools slowly at deeper sites. Feldspars are often colorless but are generally less transparent than quartz. Feldspars in soil tend to have various shapes with broken edges and some parallel cleavage surfaces within or bounding the particles.

Figure 2.13a, b show an example of a nearly un-weathered plagioclase collected from a lahar deposit from the 1991 Mt. Pinatubo eruption. In the optical micrograph (Fig. 2.13a), parallel cleavage planes can be observed weakly within the particle.

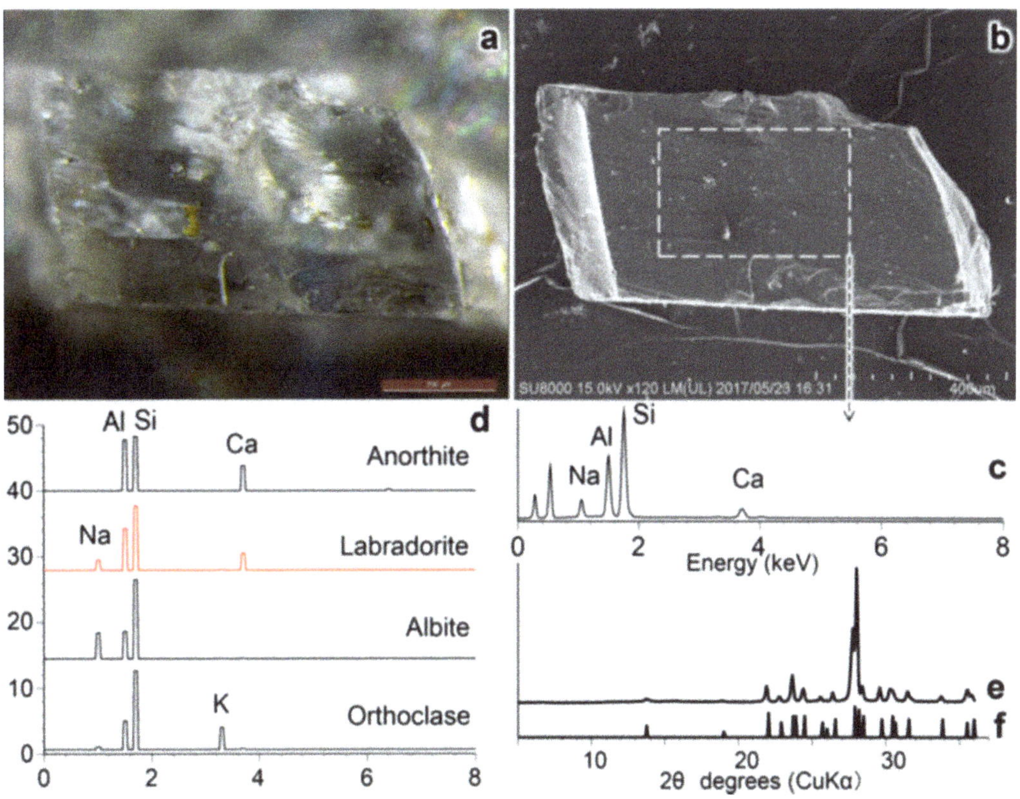

Fig. 2.13 A feldspar particle from the Mt. Pinatubo lahar deposit (1991). (**a**) Optical micrograph, (**b**) SEM image, (**c**) EDX spectrum of the dashed area in (**b**), (**d**) elemental compositions of the members having the intermediate number of Ca atoms among the 7 anorthites, the intermediate number of Ca atoms among the 10 labradorites, the intermediate number of Na atoms among the 10 albites, and the intermediate number of K atoms among the 62 K-feldspars listed by Deer et al. (2001). The powder XRD pattern of the hand-picked feldspar particles (**e**) is close to that of the natural labradorite (**f**) reported by Goodyear and Duffin (1955). The latter powder XRD pattern was obtained as a diffraction photograph, and the peak heights represent relative intensities that were estimated visually

The EDX spectrum-mimic graphs (Fig. 2.13d) show the number of cations per 32 oxygens. Four representatives were chosen from among the 7 anorthites, 10 labradorites, 10 albites, and 62 K-feldspars (sanidine, orthoclase, microcline, amazonite, etc.) listed by Deer et al. (2001). Referring to Fig. 2.13d, the EDX spectrum shown in Fig. 2.13c is close to that of labradorite, one of the plagioclase feldspars.

Plagioclase also exists as high- and low-temperature phases. The high- and low-temperature types of albite and oligoclase can be identified from a powder XRD pattern, but it gradually becomes difficult with a further increase in the ratio of anorthite (Huang 1989). The powder XRD pattern (Fig. 2.13e) appears closer to, although not completely the same as, the pattern of natural labradorite listed as No. 6 by Goodyear and Duffin (1955), low-temperature labradorite, than to the synthetic one, high-temperature type.

2.3.2.2 Silica Minerals

Among the various silica minerals in soils are quartz and cristobalite, which are grouped as tectosilicates (Table 2.1) (Drees et al. 1989; Deer et al. 2004). Mizota and Aomine (1975) reported cristobalite in the clay fraction of volcanic ash from Hokkaido, Japan.

Quartz, in particular, is found in the silt and sand fractions of many soils, although clay-sized quartz does exist in some soils affected by airborne dust. Quartz grains are colorless and transparent in many cases. There are two structure types, high- and low-temperature types, distinguishable by their XRD patterns (Brindley and Brown 1980), and the one in soils is the low-temperature type. Cracks sometimes can be found in quartz grains, possibly due to shrinkage as they cooled and converted from the high-temperature type to the low-temperature type. Conchoidal fractures are also a characteristic of quartz grains. Surface etchings due to partial dissolution are very slight.

Figure 2.14a (optical micrograph) and (b) (SEM image) show a colorless and transparent quartz grain separated from a volcanic ash deposit (1991) remobilized in a lahar from Mt. Pinatubo (Fig. 2.7). A conchoidal fracture can be observed on the left side (Fig. 2.14b). As is often the case of minerals in volcanic ash, the quartz grain is partly covered by volcanic glass. The EDX spectrum of the exposed quartz surface shows only Si (Fig. 2.14c). The powder XRD pattern (Fig. 2.14d) of a hand-picked quartz grain from the volcanic ash is identical to that of the low-temperature type of quartz (Fig. 2.14e) (Brindley and Brown 1980), although a very small reflection peak at about $2\theta = 28$ degrees, probably from plagioclase, is included in Fig. 2.14d.

All the quartz in soil is the low-temperature phase because inter-conversion between high- and low-temperature quartz occurs rapidly. The powder XRD pattern for high-temperature quartz is different from that of low-temperature quartz.

Further readings for silica minerals are Drees et al. (1989), Monger and Kelly (2002) and Deer et al. (2004).

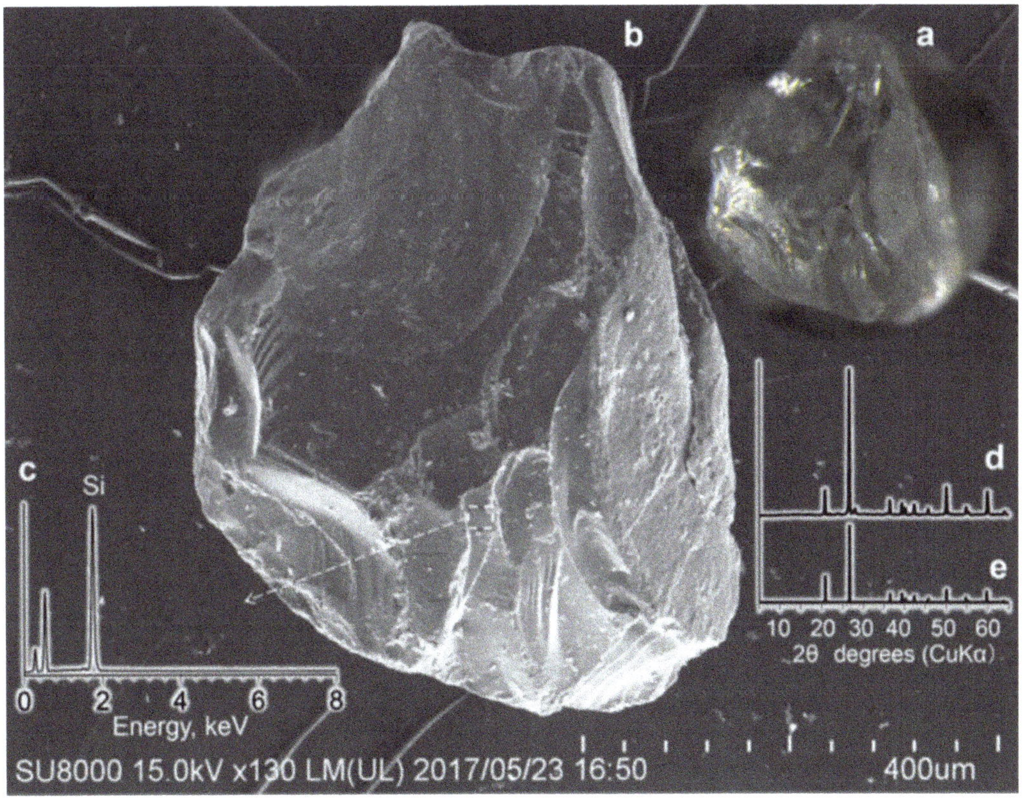

Fig. 2.14 Quartz from the 0.2–0.5 mm fraction of the 1991 Pinatubo tephra taken from the lahar deposit (Fig. 2.7). (**a**) Optical micrograph, (**b**) SEM image, (**c**) EDX spectrum acquired from the dashed area shown in (**b**), (**d**) powder XRD pattern of quartz grains, (**e**) reference powder XRD pattern (Brindley and Brown 1980)

2.4 Other Minerals in Soil

2.4.1 Titanomagnetite and Ilmenite

Iron-titanium (Fe-Ti) oxides are accessary minerals in many soils. The Fe-Ti oxides include magnetite [Fe_3O_4], titanomagnetite [(1-x)Fe_3O_4.xFe_2TiO_4] and ilmenite [$FeTiO_3$] (Allen and Hajek 1989; Milnes and Fitzpatrick 1989). Chemical compositions of these oxides are conveniently displayed on the TiO_2-FeO-1/2Fe_2O_3 ternary diagram (Butler 1992).

Figure 2.15a is an optical micrograph of magnetite particles collected from the lahar deposit from Mt. Pinatubo (Fig. 2.7) using a hand magnet, suggesting that these particles are rich in titanomagnetite. Titanomagnetite has dark color and is not transparent. Figure 2.15b is the SEM image of the dashed part of Fig. 2.15a, and the dashed square in Fig. 2.15b yielded the EDX spectrum shown in Fig. 2.15c, which suggests the elemental composition of ferrian ilmenite (Imai et al. 1996) or titanohematite (Butler 1992). An EDX spectrum with smaller amount of Ti, (Fig. 2.15d), was obtained from other part of the same particle lacking a glass coating. The powder XRD pattern of the crushed magnetic particles is shown as

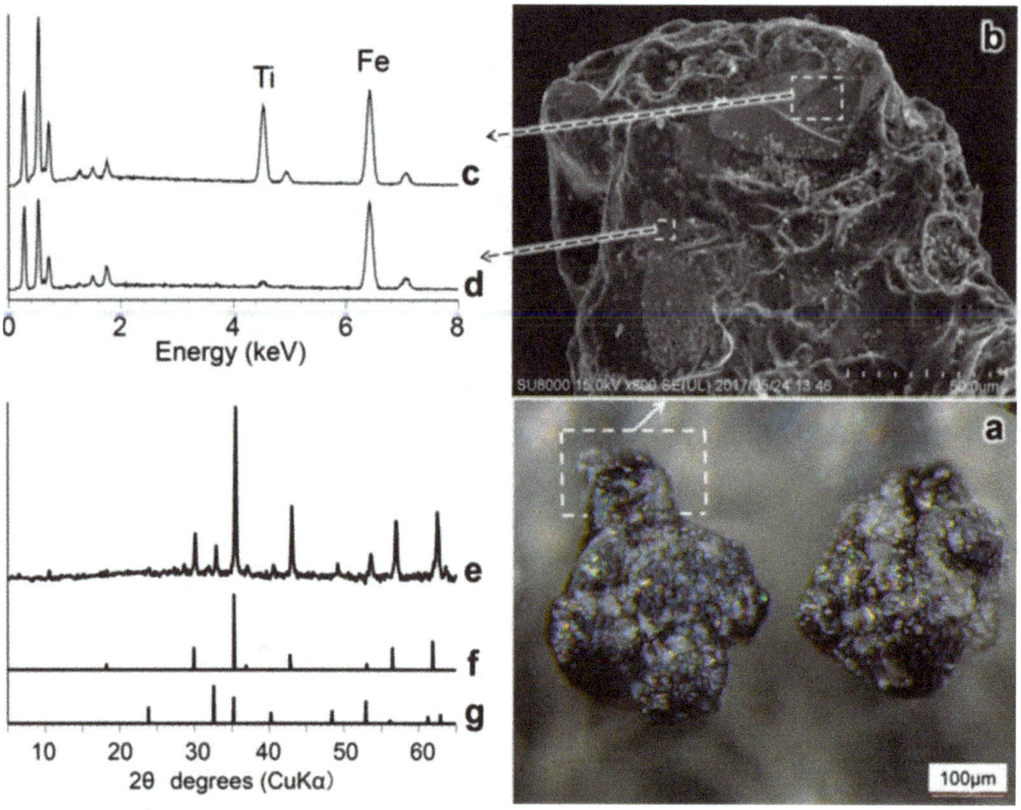

Fig. 2.15 Magnetic minerals collected from the Mt. Pinatubo lahar deposit (Fig. 2.7). (**a**) Optical micrograph, (**b**) magnified SEM image of the dashed area in (**a**), (**c** and **d**) EDX spectra of the dashed area in (**b**) and a magnetic particle with little Ti, respectively, (**e**) powder XRD pattern. Powder XRD patterns of titanomagnetite (Joint Committee on Powder Diffraction Standards No.75-1376) (**f**) and ilmenite (Brindley and Brown 1980) (**g**) are shown as references

Fig. 2.15e. It is evident that the major XRD pattern is similar to that of titanomagnetite (Joint Committee on Powder Diffraction Standards No. 75-1376) and that the XRD pattern close to ilmenite (Brindley and Brown 1980) is also included. It is frequently difficult to distinguish magnetite from titanomagnetite, and ilmenite from hematite using XRD because of their somewhat similar patterns (Milnes and Fitzpatrick 1989).

As titanomagnetite is soluble, at least partly, in oxalate solution, it affects the evaluation of Fe in soils when using oxalate extraction (Rhoten et al. 1981; Walker 1983; Shoji et al. 1987). Accordingly, in the requirements for Andisols (Soil Survey Staff 1999) and Andosols (IUSS Working Group WRB 2015), the phosphate retention percentage is used together with the oxalate extraction.

2.5 Mineral Samples in Soil Derived from a Weathered Granitic Rock

Relatively fresh or un-weathered primary minerals were introduced in the previous sections. However, the minerals in soil environments are physically and chemically altered to varying degrees under the activity of biota. In this section, a soil formed on

Cretaceous granitoids (Kon et al. 2015) is exemplified. The soil is located at the dam lakeshore of the Utsushi–Gawa River (Figs. 6.17 and 6.24), bottom of a valley slope in the central Abukuma Plateau, northeastern Japan Arc. Although the age of the bedrock is old, the soil developed at the bottom of the valley slope can be much younger than the bedrock due to subsequent continuous erosion. This soil is a shallow Udept with A, Bw, and C horizons under a bamboo forest (Fig. 2.16a, b). The 2–0.2 mm fraction of the Bw horizon includes quartz, feldspar, biotite, hornblende, and others showing different intensities of weathering (Fig. 2.16c).

Among the mineral particles in Fig. 2.16c, quartz has a high resistance to weathering. Figure 2.16d shows an optical micrograph of a particle that consists mostly of quartz. Although a crack occurs at the arrowed site of the SEM image (Fig. 2.16e), it may have formed by shrinkage during the phase transformation from high- to low-temperature quartz. Although conchoidal fractures, which are characteristic of quartz, are found on the surface (Fig. 2.16g), almost no etching due to weathering can be seen.

The feldspar particle in Fig. 2.17a has a whitish color and is not transparent. Similar particles can be found in Fig. 2.16c. According to the SEM image (Fig. 2.17b) and the magnified SEM image (Fig. 2.17c), surface etching and opening of cleavage planes are evident due to weathering.

Fig. 2.16 Coarse sand fraction of the Bw horizon of soil sampled from the dam lakeshore of the Utsushi–Gawa River, Japan. (**a**) Vegetation, (**b**) soil profile, (**c**) 2–0.2 mm fraction of Bw horizon, (**d**) selected quartz particle, (**e**) SEM image of (**d**), (**f**) EDX spectrum obtained from the larger dashed area in (**e**), (**g**) magnified SEM image of the smaller dashed area indicated in (**e**)

Fig. 2.17 Weathered feldspar from the 2–0.2 mm fraction (Fig. 2.16c). (**a**) Optical micrograph, (**b**) SEM image, (**c**) magnified SEM image of the dashed area in (**b**), (**d** and **e**) EDX spectra of the dashed areas in (**c**). The EDX spectra (**d**) and (**e**) suggest plagioclase and K-feldspar, respectively

Another property found in feldspar is complexity or microtexture. For example, two EDX spectra, (d) and (e), were obtained from the selected areas in Fig. 2.17c. The EDX spectra of Fig. 2.17d, e corresponds to those of plagioclase and orthoclase in Fig. 2.13d, respectively.

Using a polished section (Fig. 2.18a), the combination of different minerals and microtextures in a gravel particle can be observed clearly by using SEM-EDX. The polished section was prepared by embedding gravel particles in a resin and then cutting and polishing (see Sect. 1.4). The arrowed particle from Fig. 2.18a was selected. The optical micrograph (Fig. 2.18b) shows that this gravel particle has several different parts, weakly transparent, cloudy, and dark. Brown, fine-textured soil materials partly surround the gravel particle.

The element maps in Fig. 2.18 show that this gravel particle consists of several different minerals, more than are apparent from the optical micrograph (Fig. 2.18b). Looking at the Si element map, there are more than three levels of different Si concentration areas, suggesting different minerals. From these element maps, at least six different mineral areas can be identified, as listed below (numbered in the Si map of Fig. 2.18):

(1) An area with high Mg and Fe concentrations and medium Ca concentration.
(2) Areas with the highest Si concentration and near-zero concentrations of other elements.

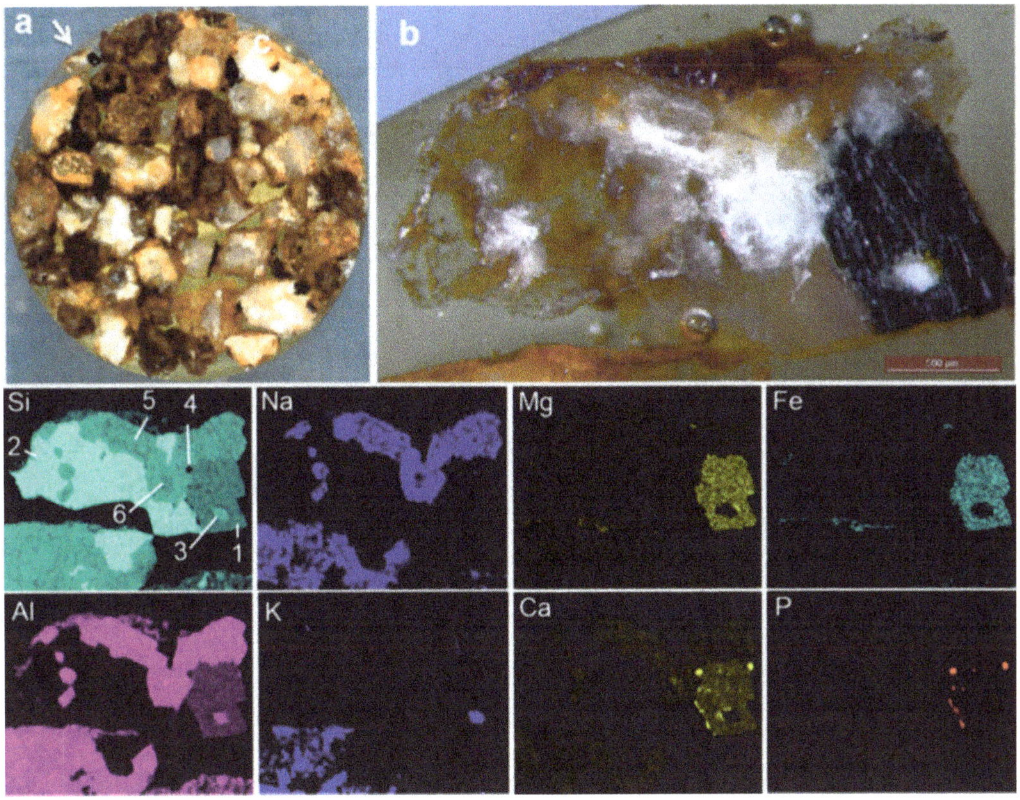

Fig. 2.18 Gravel particles of the Bw horizon (Fig. 2.16b). (**a**) Polished section, (**b**) optical micrograph of the particle arrowed in (**a**), and element maps of Si, Al, Na, K, Mg, Ca, Fe, and P obtained by SEM-EDX

(3) Areas with high K concentration and almost no Na and Ca atoms.
(4) Areas with the highest Ca and P concentrations.
(5) Areas with high Na and Al concentrations and low Ca concentration.
(6) A small area with the highest Al concentration, near-zero Na and K concentrations, and medium Ca concentration.

The three EDX spectra (Fig. 2.19b–d) represent the areas (b), (c), and (d), respectively, shown in Fig. 2.19a. A comparison of the EDX spectra shown in Sects. 2.3 and 2.4 with those of Fig. 2.19b–d suggest hornblende (the area (1)), quartz (the areas (2)), and K-feldspar (the areas (3)), respectively. The areas (4) correspond to apatite, which is discussed in Sect. 6.3.

The small area (6) of Fig. 2.18, also selected as Fig. 2.19e, shows the highest Al concentration, almost no Na, no K, and a medium-level concentration of Ca. A further complex pattern can be observed in the magnified SEM image (Fig. 2.20). The EDX spectrum obtained from the small area having medium Ca concentration (Fig. 2.20b) suggests anorthite (Fig. 2.13d). The areas around the anorthite are close to Na-rich plagioclase, or oligoclase (the areas (5)) (Fig. 2.20a). When solid solutions of high-temperature feldspars cool, they tend to separate into their end-members and form a complex microtexture of the end-members (Huang 1989).

Thus, the gravel particle resembles a rock fragment having complex minerals. Furthermore, the microtextures of the feldspars might have resulted from the

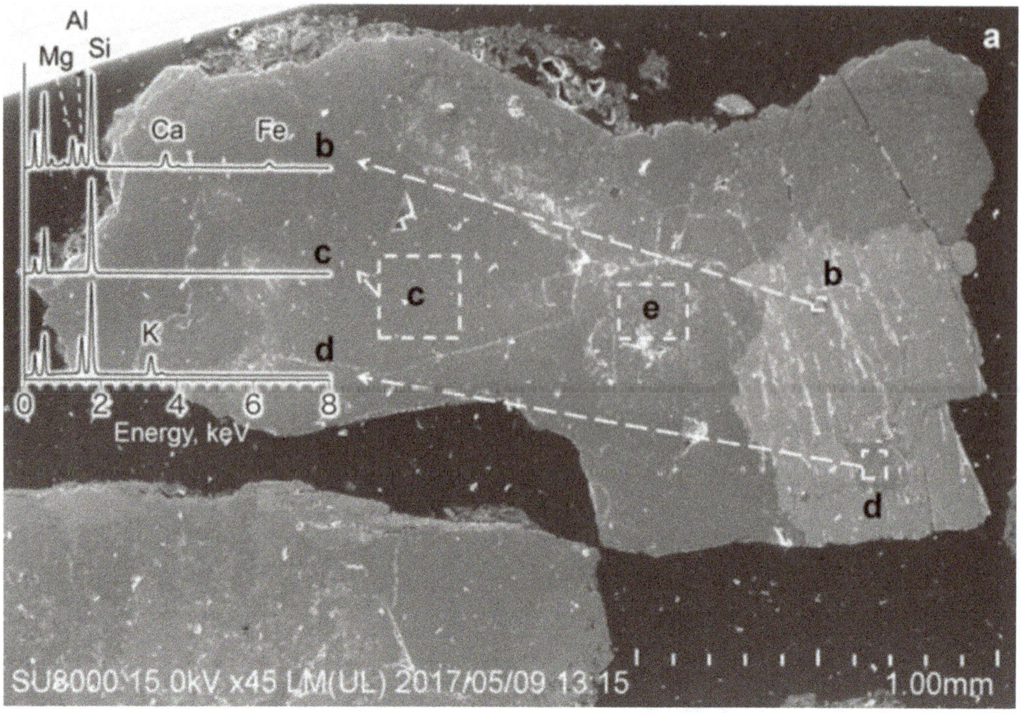

Fig. 2.19 Polished section of a gravel particle (Fig. 2.18b). (**a**) SEM image, (**b**, **c**, and **d**) EDX spectra of the dashed areas (**b**), (**c**), and (**d**), respectively. The dashed area (**e**) includes a relatively high Ca area

Fig. 2.20 Magnified SEM image of Fig. 2.19e and EDX spectra (**a** and **b**) of dashed areas **a** and **b**, respectively

Fig. 2.21 A weathered biotite particle separated from the 2–0.2 mm fraction (Fig. 2.16c). (**a** and **b**) Optical micrographs of a plane site and an edge site, respectively, (**c**) EDX spectra of the selected areas in (**a**) and (**b**), (**d**) magnified SEM image of the edge site of (**b**)

conversion of high- to low-temperature phases during the cooling process. A single feldspar particle can be composed of different feldspars, as suggested by Fig. 2.17c.

Biotites tend to weather much more easily than muscovite and many other minerals. The coarse fraction (2–0.2 mm) exemplified in Fig. 2.9c includes many weathered biotite particles, as well as muscovite particles that show only weak weathering. The coarse sand fraction of Fig. 2.16c also contains many weathered biotite particles. Weathered biotite in soil shows mostly brown or golden colors, although the hexagonal platy properties of the mineral may partly or wholly remain (Fig. 2.21a). The originally flat cleavage face becomes wavy, and the edges of the particle exfoliate along cleavage (Fig. 2.21b, d). The remaining K is distributed non-uniformly (see EDX spectra in Fig. 2.21c). Hydrobiotite, vermiculite, and halloysite have been detected in weathered biotite particles (Masui 1954).

Hornblende particles also occur in the sand fraction (Fig. 2.16c). Although the hornblende particles maintain their prismatic appearance, they are easily broken into finer fragments when handled by tweezers, indicating that they are also partly weathered.

The weathering intensity of the primary minerals in the sand fraction of the exemplified granitic soil is variable depending on the mineral species. Further variation may be added by climatic conditions, redox conditions, age of the soil, and other environmental factors. Thus, partially weathered sand-size minerals may exchange or fix ions. As will be shown in Sect. 6.2, weathered biotite particles in the sand fraction fix radiocesium. Other possible functions supporting plant growth are the release and tentative retention of nutrient elements, water, etc., although the capacity of these functions of the coarse fraction is not as high as that of clays and humus.

Further readings are Huang (1989) and Deer et al. (2013)

References

Allen BL, Hajek BF (1989) Mineral occurrence in soil environments. In: Dixon JB, Weed SB (eds) (Co-eds) minerals in soil environments, 2nd edn. Soil Science Society of America, Madison, pp 199–278

Brindley GW, Brown G (1980) Crystal structures of clay minerals and their X-ray indentification. Mineralogical society monograph no.5. Mineralogical Society, London

Butler RF (1992) Ferromagnetic minerals. In: Paleomagnetism: Magnetic domains to geologic terranes, Blackwell Scientific Publications, pp16–30

Deer WA, Howie RA, Zussman J (1997a) Rock-forming minerals, Orthosilicates, 2nd edn. The geological society, London

Deer WA, Howie RA, Zussman J (1997b) Rock-forming minerals, single-chain silicates, 2rd edn. The geological society, London

Deer WA, Howie RA, Zussman J (1997c) Rock-forming minerals, double-chain silicates, 2rd edn. The geological society, London

Deer WA, Howie RA, Zussman J (2001) Rock-forming minerals, framework silicates: feldspars, 2nd edn. The geological society, London

Deer WA, Howie RA, Zussman J (2004) Rock-forming minerals, framework silicates: silica minerals, feldspathoids and the zeolites, 2nd edn. The geological society, London

Deer WA, Howie RA, Zussman J (2013) An introduction to the rock-forming minerals, 3rd edn. Mineralogical society, London

Drees LR, Wilding LP, Smeck NE, Senkayi AL (1989) Silica in soils: quartz and disordered silica polymers. In: Dixon JB, Weed SB (eds) (Co-eds) minerals in soil environments, 2nd edn. Soil Science Society of America, Madison, pp 913–974

Fanning DS, Keramidas VZ, El-Desoky MA (1989) Micas. In: Dixon JB, Weed SB (eds) (Co-eds) minerals in soil environments, 2nd edn. Soil Science Society of America, Madison, pp 551–634

Fleet ME (2003) Rock-forming minerals, micas, 2nd edn. The geological society, London

Glossary of Soil Science Terms Committee (2008) Glossary of soil science terms. Soil Science Society of America, Inc, Madison

Goodyear J, Duffin WJ (1955) The identification and determination of plagioclase feldspars by the X-ray powder method. Miner Mag J Miner Soc 30:306–326

Huang PM (1989) Feldspars, olivines, pyroxenes, and amphiboles. In: Dixon JB, Weed SB (eds) (Co-eds) minerals in soil environments, 2nd edn. Soil Science Society of America, Madison, pp 975–1050

Imai A, Listanco EL, Fujii T (1996) Highly oxidized and sulfur-rich dacitic magma of Mount Pinatubo: Implication for metallogenesis of porphyry copper mineralization in the western Luzon arc. In: CG Newhall and RS Punongbayan (eds) Fire and mud, Eruptions and lahars of Mount Pinatubo, Philippines, Philippine Institute of Volcanology and Seismology, Quezon city, University of Washington Press, Seattle and London, pp 865–874

IUSS Working Group WRB (2015) World reference base for soil resources 2014. International soil classification system for naming soils and creating lengends for soil maps, update 2015, World soil resources reports no.106. FAO, Rome

Joint Committee on Powder Diffraction Standards (1986) Mineral powder diffraction file: data book. International Center for Diffraction Data, Swarthmore

Kon Y, Morita S, Takagi T (2015) Spatial U-Pb age distribution of plutonic rocks in the central Abukuma plateau, northeastern Japan Arc. J Miner Petr Sci 110:145–149

Masui J (1954) Mineralogical studies on the soil genesis (2): the clay minerals in the soil derived from granodiorite at Ogoe, Fukushima Prefecture (I). J Miner Petrol Sci 38:165–176 (In Japanese with English abstract)

Milnes AR, Fitzpatrick RW (1989) Titanium and zirconium minerals. In: Dixon JB, Weed SB (eds) (Co-eds) minerals in soil environments, 2nd edn. Soil Science Society of America, Madison, pp 1131–1205

Mizota C, Aomina S (1975) Clay mineralogy of some volcanic ash soils in which cristobalite predominates. Soil Sci Plant Nutr 21:327–335

Mizuno N, Amano Y, Mizuno T, Nanzyo M (2008) Changes in the heavy minerals content of Tarumae-a tephra with distance from the source volcano and its effect on the element concentration of the tephra. Soil Sci Plant Nutr 54:839–845

Monger HC, Kelly EF (2002) Silica minerals. In: Dixon JB, Schulze DG (eds) (Co-eds) soil mineralogy with environmental applications. Soil Science Society of America, Inc, Madison, pp 611–636

Nanzyo M, Nakamaru Y, Yamasaki S, Samonte HP (1999) Effect of reducing conditions on the weathering of Fe^{3+}-rich biotite in the new lahar depsoit from Mt. Pinatubo, Philippines. Soil Sci 164:206–214

Rhoten FE, Bigham JM, Norton LD, Smeck ME (1981) Contribution of magnetite to oxalate-extractable iron in soils and sediments from the Maumee River basin of Ohio. Soil Sci Soc Am J 45:645–649

Shoji S, Ito T, Saigusa M (1987) Andisol-Entisol transition problem. Pedologist 31:171–175

Skinner BJ, Mruck B (2011) The rock cycle. In: The blue planet an introduction to earth system science, 3rd edn. Wiley, New York, pp 64–71

Soil Survey Staff (1999) Soil taxonomy, a basic system of soil classification of making and interpreting soils surveys, USDA-NRCS, agriculture handbook no. 436, U.S. Government Printing Office, Washington, DC

Thompson M, Ukrainczyk L (2002) Micas. In: Dixon JB, Schulze DG (eds) (Co-eds) soil mineralogy with environmental applications. Soil Science Society of America, Inc, Madison, pp 431–466

Walker (1983) The effect of magnetite on oxalate- and dithionite-extractable iron. Soil Sci Soc Am J 47:1022–1026

Wolfe EW, Hoblitt RP (1996) Overview of the eruption. In: Newhall CG, Punongbayan RS (eds) . Fire and mud, Eruptions and lahars of Mount Pinatubo, Philippines, Plilippine Institute of Volcanology and Seismology/University of Washington Press, Quezon city/Seattle/London, pp 3–20

Wyllie PJ (1971) The structure, petrology, and composition of the earth' crust. In: The dynamic earth: textbook in geoscience. Wiley, New York, pp 139–155

Chapter 3
Secondary Minerals

Abstract This chapter introduces secondary minerals such as clay minerals, hydroxides and oxides in soils. Clay minerals, which are major secondary minerals in soils, are phyllosilicates that have 1:1 or 2:1 type layers. The 1:1 type minerals are kaolinite and halloysite. The 2:1 type minerals are smectite, vermiculite, micaceous minerals, and chlorite. Gibbsite and manganese oxides are introduced as examples of hydroxides and oxides, respectively. This chapter also constitutes the basic part of treatment of the inorganic constituents in soils. Secondary minerals have an effect on the chemical, physical, and biological functions of soils. As the size of secondary minerals is approximately 2 μm or less, electron micrographs and X-ray diffraction (XRD) are effective for characterizing these minerals. Schematic diagrams are used to interpret the chemical structure of phyllosilicate clay minerals.

3.1 Introduction

The secondary minerals in soil are minerals that are stabilized under various soil environments. The major secondary phyllosilicates in soil are summarized in Table 3.1 with their idealized chemical formulae (Deer et al. 2013). Formula units in association with $O_5(OH)_4$ and $O_{10}(OH)_2$ are also used for 1:1 and 2:1 type phyllosilicate minerals, respectively (Kodama 2012). The major oxides, hydroxides, and other relevant minerals in soil (Kampf et al. 2012) are summarized in Table 3.2. This table includes minerals described in other chapters. Other non-crystalline secondary and primary inorganic materials in soil are described in Chap. 4. These minerals and materials in soil can be altered further if environmental conditions change.

Environmental factors that affect the behavior of secondary minerals include water, temperature, redox condition, biological activities, and time (Churchman and Lowe 2012). In addition, human activities affect secondary minerals and materials through agricultural soil management, civil engineering, and other activities. As examples, the dry–wet cycle affects the shrink–swell behavior of smectitic soils, and the exchangeable cation composition of secondary minerals is easily affected by application of a fertilizer, inundation of seawater, etc. Changes in

M. Nanzyo, H. Kanno, *Inorganic Constituents in Soil*,
https://doi.org/10.1007/978-981-13-1214-4_3

Table 3.1 Major secondary phyllosilicate minerals in soil

Secondary phyllosilicate	Chemical formula
Kaolinite	$Al_4Si_4O_{10}(OH)_8$
Halloysite (1.0 nm)	$Al_4Si_4O_{10}(OH)_8 \cdot 4H_2O$
Halloysite (0.7 nm)	$Al_4Si_4O_{10}(OH)_8$
Montmorillonite	$M_{0.67}Si_8(Al_{3.33}Mg_{0.67})O_{20}(OH)_4 \cdot nH_2O$
Beidellite	$M_{0.67}(Si_{7.33}Al_{0.67})Al_4O_{20}(OH)_4 \cdot nH_2O$
Nontronite	$M_{0.67}(Si_{7.33}Al_{0.67})Fe_4O_{20}(OH)_4 \cdot nH_2O$
Vermiculite	$(Mg,Ca)_{0.6–0.9}(Al,Si)_8(Mg,Fe^{3+},Al)_6O_{20}(OH)_4 \cdot nH_2O$
Illite	$K_{1.5–1.0}(Si_{6.5–7.0}Al_{1.5–1.0})\ Al_4O_{20}(OH)_4$
Mg chlorite	$Mg_4Al_2(OH)_{12}(Si_6Al_2)Mg_6O_{20}(OH)_4$

M: Exchangeable cation

Table 3.2 Oxides, hydroxides, and other related inorganic constituents in soil

	Chemical formula
Brucite	$Mg(OH)_2$
Gibbsite	$Al(OH)_3$
Hematite	Fe_2O_3
Goethite	α-FeOOH
Lepidocrosite	γ-FeOOH
Ferrihydrite	$Fe_5O_7(OH) \cdot 4H_2O$
Siderite	$FeCO_3$
Lithiophorite	$(Al, Li)MnO_2(OH)_2$
Birnesite	$(Na, Ca)Mn_7O_{14} \cdot 2.8H_2O$
Apatite	$Ca_5(PO_4)_3(F, Cl, OH) \cdot H_2O$
Brushite	$CaHPO_4 \cdot 2H_2O$
Struvite	$MgNH_4PO_4 \cdot 6H_2O$
NH_4-taranakite	$Al_5(NH_4)_3H_6(PO_4)_8 \cdot 18H_2O$
Vivianite	$Fe_3(PO_4)_2 \cdot 8H_2O$
Natrojarosite	$NaFe_3(OH)_6(SO_4)_2$
Gypsum	$CaSO_4 \cdot 2H_2O$
Calcite	$CaCO_3$
Dolomite	$CaMg(CO_3)_2$

redox conditions affect the dissolution and precipitation of iron-bearing minerals and materials. The dissolution and precipitation of iron-bearing minerals and materials is introduced in Chap. 5, and sodification due to seawater inundation is introduced in Sect. 6.2.

Phyllosilicates are the major secondary minerals in soils, and the properties of soils depend on the composition and structure of phyllosilicates. Hence, this chapter begins with the structures of phyllosilicates and then discusses the relationship between the structure and function of phyllosilicates. For identification of phyllosilicates in soil, the changes in basal spacing that occur with changes in exchangeable cations are studied. The basal spacing can be determined with XRD, and changes in the XRD pattern with changes in exchangeable cations (Harris and

White 2008) are shown for several phyllosilicates, with the expectation that these will be helpful for understanding the properties of the phyllosilicates (Molloy and Kerr 1961) as well as scanning electron microscope (SEM) and transmission electron microscope (TEM) images (Sudo and Shimoda 1978).

3.2 Construction of Layer Aluminosilicate Models

The major phyllosilicates of the clay fraction in soil are aluminosilicates. The aluminosilicates are constructed from the phyllosilicate sheet (Fig. 2.3) and the Al octahedral sheet. The octahedral sites of the phyllosilicates may be wholly or partly occupied by Mg or Fe. This section begins with the construction of the Mg octahedral sheet, brucite, followed by the Al octahedral sheet, and the 1:1 type aluminosilicate. Attention is paid to Al–O–Si bonding and the arrangement between the Al octahedral sheet and the Si tetrahedrons of the phyllosilicate sheet. The structure of 1:1 type aluminosilicates to 2:1 types is extended similarly.

3.2.1 Brucite Sheet and Gibbsite Sheet

The construction of the brucite sheet, $Mg(OH)_2$, is shown in Fig. 3.1. On the distant arrangement of blue balls representing OH groups (back OH), small orange-colored balls (Mg) are placed (Fig. 3.1a). The size of the balls roughly indicates the relative

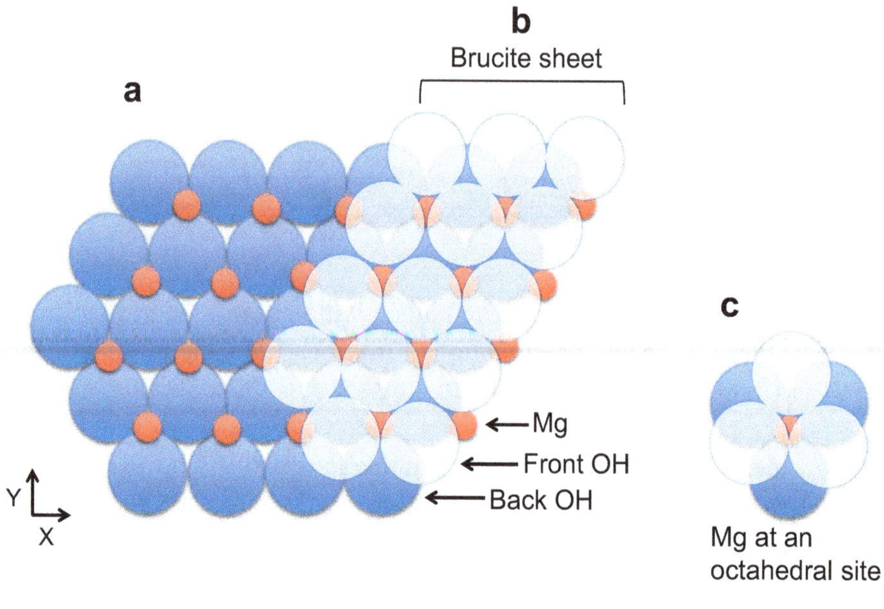

Fig. 3.1 Schematic of a brucite sheet. (**a**) Placement of Mg on back OH groups, (**b**) placement of front OH groups on (**a**) to form a brucite sheet, (**c**) location of Mg at the center of an octahedral site

size of each ion. The procedure of building the sheet is to place three Mg balls alternately among six possible sites around the back OH balls, as shown in Fig. 3.1a. As a result, three Mg sites and three vacant sites are formed around each back OH ball. The front OH balls are placed on the three vacant sites, and the Mg ion is located at the center of an octahedron formed by six OH balls, as shown in Fig. 3.1c. Thus, a brucite sheet is completed (Fig. 3.1b).

3.2.2 *Construction of Gibbsite Sheet and 1:1 Layer Aluminosilicate*

In the brucite sheet shown in Fig. 3.1, all of the octahedral sites are filled with a Mg ion. In the aluminosilicates, Al ions occupy the octahedral sites instead of Mg ions. Because an Al ion has three positive charges and a Mg ion has two positive charges, one-third of the octahedral sites must be vacant for electroneutrality (Fig. 3.2a). In this way, after two-thirds of the Mg ion sites of the brucite sheet are replaced by Al ions and the remaining Mg ions are removed to make vacant octahedral sites, a gibbsite sheet is formed (Fig. 3.2b).

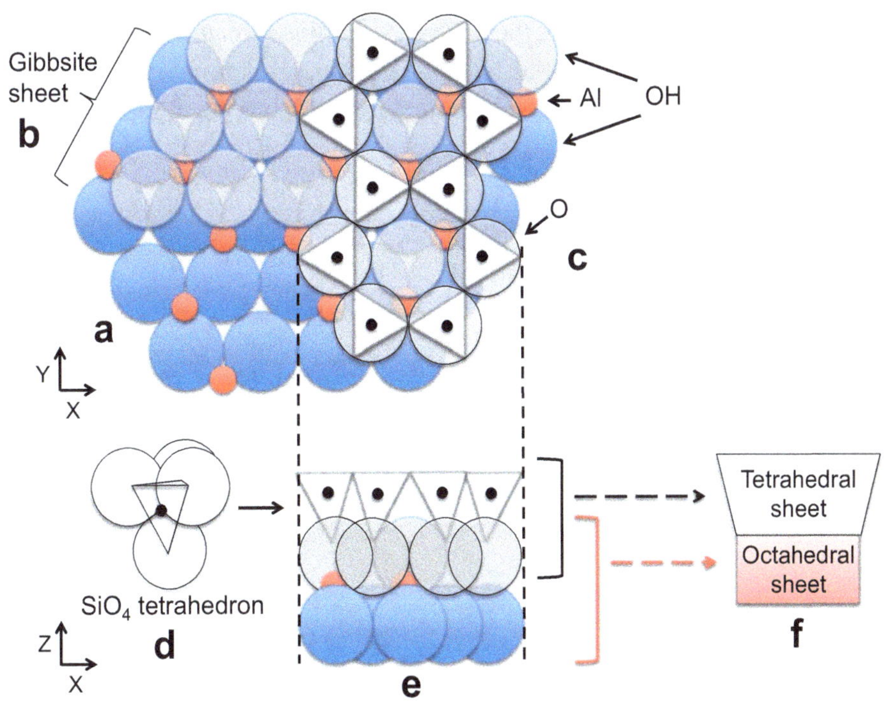

Fig. 3.2 An assembly diagram of a 1:1 dioctahedral aluminosilicate sheet (**a** → **b** → **c**). (**a**) Placement of Al on back OH groups, (**b**) placement of front OH groups to form a gibbsite sheet, (**c**) placement of SiO_4 tetrahedrons onto the front OH groups of the gibbsite sheet, (**d**) location of Si at the center of a tetrahedral site. The SiO_4 group is shown by tetrahedrons (see (**d**)) in (**c**) and (**e**). A horizontal sectional view of (**c**) is shown as (**e**) and further simplification of (**e**) is shown as (**f**)

Connecting a phyllosilicate sheet onto the front OH balls produces a dioctahedral aluminosilicate sheet, as shown in Fig. 3.2c, where Si ions are placed (small black balls) in the center of SiO_4 tetrahedrons (Fig. 3.2d). The reaction of the gibbsite sheet and the phyllosilicate sheet can be formulated simply as

$(OH)_2$Al-Front OH (gibbsite, Fig. 3.2(b))
$+ SiO_{1.5}O^-H^+$ (one fourth of phyllosilicate unit cell coordinated with H^+, Fig. 2.3)
$= (OH)_2Al{-}O{-}SiO_{1.5} + H_2O$

where an Al–O–Si bond is formed with the release of a molecule of H_2O (Fig. 3.2e). A front OH group remains at the center of the Si tetrahedron ring (Fig. 3.2c). Further schematic simplification of the tetrahedral and octahedral sheets in Fig. 3.2e gives Fig. 3.2f.

3.2.3 Major Layer Aluminosilicates in Soil

Using the simplification of the tetrahedral sheet and octahedral sheet shown as Fig. 3.2f, the major layer aluminosilicates in soil are represented by six different aluminosilicates in Fig. 3.3. These six layer aluminosilicates can be grouped as 1:1 and 2:1 types. The 1:1 type aluminosilicates have a stacking of one tetrahedral layer and one octahedral layer, whereas the 2:1 type aluminosilicates have an octahedral layer sandwiched by two tetrahedral layers. The major 1:1 type aluminosilicates in soil are kaolinite and halloysite. Halloysite (1.0 nm) has an interlayer of water,

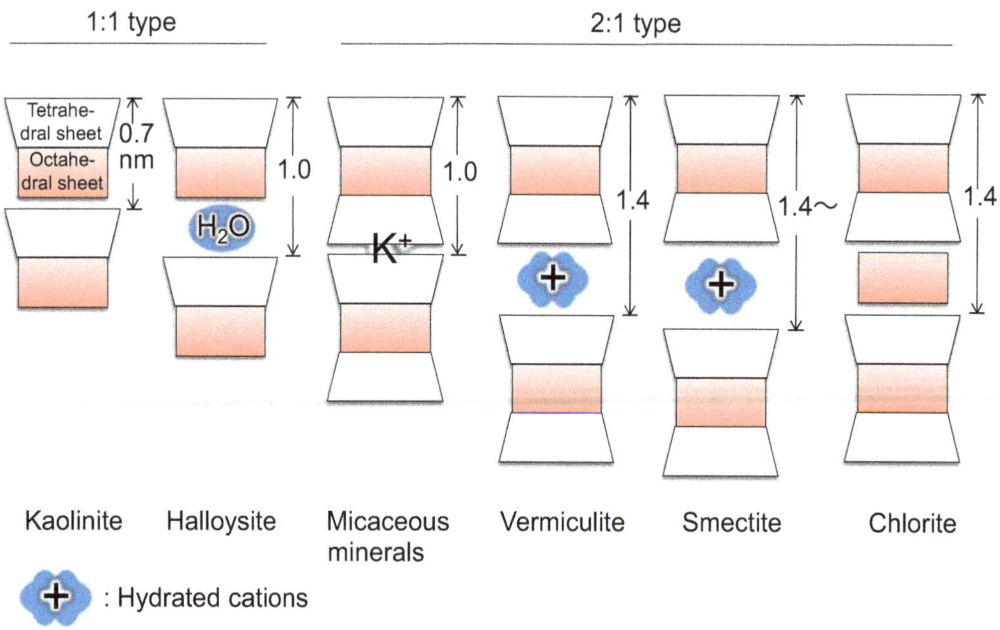

Fig. 3.3 Schematic of phyllosilicate clay minerals. Arrows show basal spacings

whereas halloysite (0.7 nm) does not. The major 2:1 type aluminosilicates are smectite, vermiculite, micaceous minerals, and chlorite.

In layer aluminosilicates, the same unit layer is stacked repeatedly. The distance between one basal plane and the next one, as shown by arrows in Fig. 3.3, is called the basal spacing. The difference in the two halloysites is indicated by the difference in the basal spacing as halloysite (0.7 nm) and halloysite (1.0 nm). The differences among the 2:1 types are characterized by the difference in the basal spacing and changes in the basal spacing that occur with several treatments. The basal spacing of these aluminosilicates is determined by XRD, and the changes in the basal spacing with several treatments (Mg saturation and glyceration, Mg saturation, K saturation, heating at 300 and 550 °C after K saturation, etc.) are used to identify the aluminosilicate clays in soil. As XRD is the most effective tool for identifying the clay mineral composition of soil, XRD patterns are used along with TEM and SEM to introduce clay minerals in the following sections.

As shown in Figs. 3.1 and 3.2, there are two types of octahedral layers. One is the brucite type in which all octahedral sites are occupied by a divalent cation, which is called the trioctahedral type. The other is the gibbsite type, called the dioctahedral type, in which two-thirds of the octahedral sites are occupied by a trivalent cation.

In the secondary minerals in soil, cations in the tetrahedral and octahedral sites are replaced by other cations with similar size and lower valence. This phenomenon is called isomorphous substitution. For example, at the tetrahedral sites, Si^{4+} is partly replaced by Al^{3+}, and at the octahedral sites, Al^{3+} is partly replaced by Mg^{2+} or Fe^{2+}. Although the isomorphous substitution does not affect the crystal structure very much, the number of positive charges decreases and a surplus of negative charge occurs in the aluminosilicate layer. The surplus of negative charge is neutralized by adsorption of cations, called exchangeable cations.

3.2.3.1 1:1 Type Minerals

The 1:1 type minerals include kaolinite and halloysite. A group name of these minerals is "kaolin minerals." Hydrogen bonds connect the 1:1 layers in kaolinite. In the idealized formula (Table 3.1), there is no isomorphic substitution.

Figure 3.4 shows a reference kaolinite sample, kaolinite No. 9 of the American Petroleum Institute (A.P.I.) reference clay minerals. The thin-to-thick platy and partially hexagonal properties of kaolinite minerals can be seen in both the TEM image (Fig. 3.4a) and the SEM image (Fig. 3.4b). XRD patterns of oriented samples show a strong diffraction peak at 0.7 nm, a basal spacing of kaolinite (Fig. 3.3). Although the basal spacing of kaolinite is not affected by the four treatments (Mg saturation and glyceration, Mg saturation, K saturation, and heating at 300 °C after K saturation), kaolin minerals are converted to an amorphous state by heating at 550 °C, and the diffraction peak at 0.7 nm disappears (Fig. 3.4c).

Kaolinite is one of the most frequently found clay minerals in soil. Figure 3.5 shows an example from the Bt2 horizon of an Ultic Palexeralf used for a vineyard in Nuble Province, Chile. The TEM image (Fig. 3.5c) and SEM image (Fig. 3.5d) show

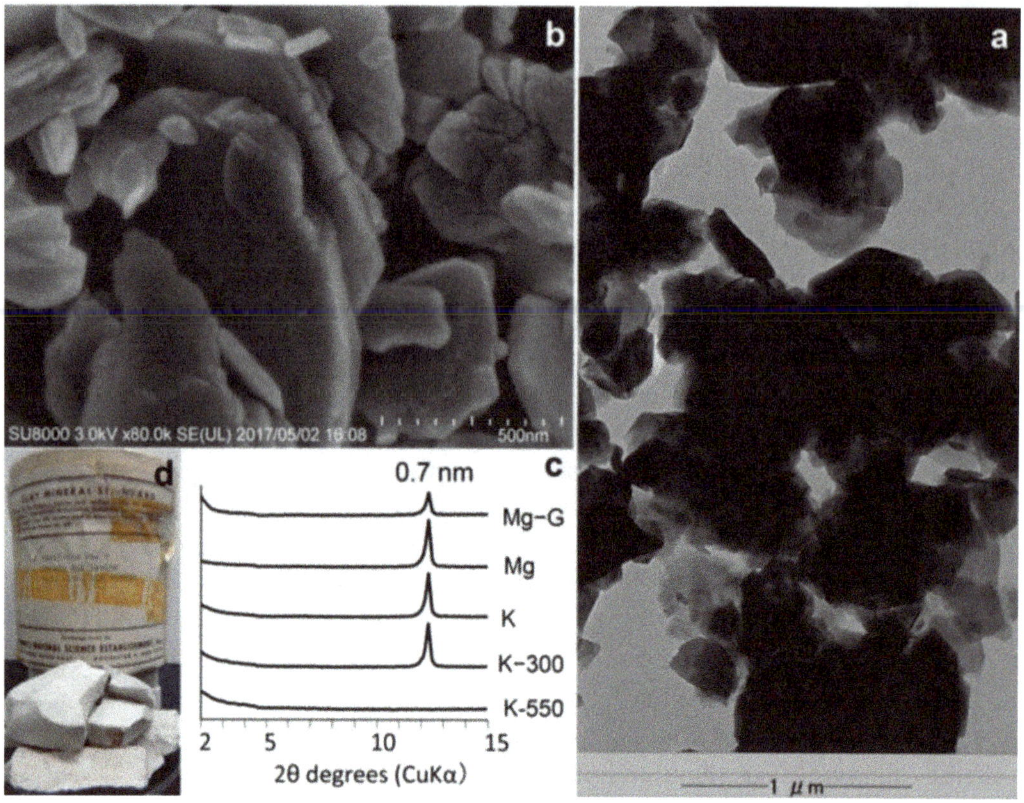

Fig. 3.4 Reference sample of kaolinite. (**a**) TEM image, (**b**) SEM image, and (**c**) XRD patterns with five treatments (Mg saturation with glyceration, Mg saturation, K saturation, heating at 300 °C after K saturation, and heating at 550 °C after K saturation) of the oriented clay fraction, (**d**) dried clods of the reference kaolinite

smaller and more sub-round platy particles than those in Fig. 3.4. The XRD patterns of the kaolinite in Fig. 3.5e show the major diffraction peak at 0.7 nm, the basal spacing of kaolinite, and are basically the same as those in Fig. 3.4c. However, the 0.7 nm diffraction peak is broader than the peak in Fig. 3.4c, which suggests lower crystallinity of the kaolinite in Fig. 3.5 compared with the kaolinite in Fig. 3.4. A small amount of 2:1 clay mineral, suggested by the diffraction peaks at 1.4 nm (Mg saturation) and 1.0 nm (K saturation, heating at 300 °C after K saturation, and heating at 550 °C after K saturation) is also indicated in Fig. 3.5e.

Kaolinite is the major clay mineral of Oxisols, Ultisols, and other soils. The cation exchange capacity (CEC) of kaolinite is typically less than 10 $cmol_c\ kg^{-1}$, and it characterizes the oxic horizon and Oxisols in the Soil Taxonomy of the United States Department of Agriculture (Soil Survey Staff 1999). Weatherable minerals in Oxisols are so limited that Oxisols are considered highly weathered soils. Thus, kaolinite is the most weathered member of the layer aluminosilicates in soil. Oxisols are distributed in central Africa and Brazil with tropical rainforest climate. They occupy about 12% of the world's soils.

Halloysite, the other major member of the 1:1 type minerals, is found in many soils as a major or minor component of the clay fraction. Halloysite also occurs in

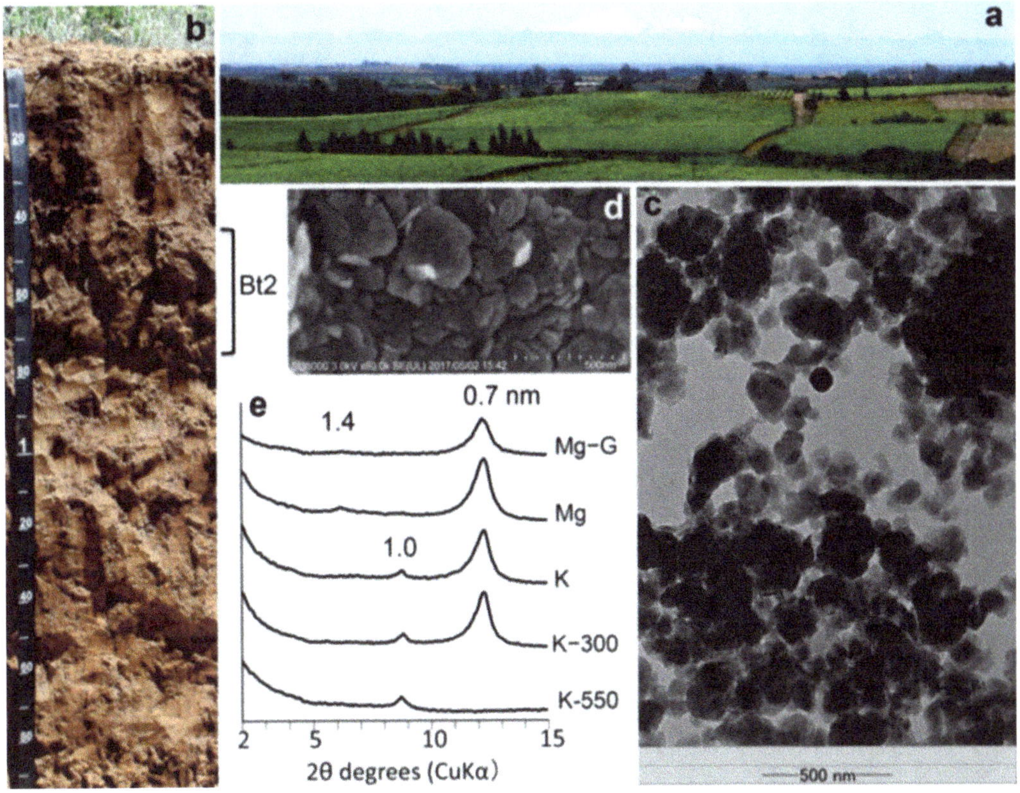

Fig. 3.5 Kaolinite in soil. (**a**) Landscape, (**b**) soil profile with kaolinite-rich soil horizons. A kaolinite-rich clay fraction was prepared from the Bt2 horizon (**b**). (**c**) TEM image, (**d**) SEM image, (**e**) XRD patterns with five treatments (Mg saturation with glyceration, Mg saturation, K saturation, heating at 300 °C after K saturation, and heating at 550 °C after K saturation) of the dithionite–citrate–bicarbonate treated and oriented clay fraction

Andisols under semi-dry climates and in the lower horizons of Andisols under humid climate. Figure 3.6 shows an example of a halloysite-containing, partially weathered volcanic ash in the lower horizon (2C2) of a soil in Aizu, Japan. A possible reason for the halloysite formation from volcanic ash is the poor drainage of this area.

The morphological and chemical properties of halloysite are diverse. Halloysite shows thin platy, curved tubular, and spherically curved properties under TEM observation (Fig. 3.6b). There are two types of halloysite, halloysite (1.0 nm) and halloysite (0.7 nm). The values in parentheses indicate their basal spacing. XRD patterns are effective for distinguishing the two types of halloysite. Glyceration causes its basal spacing to expand to 1.1 nm, and the basal spacing of halloysite (1.0 nm) decreases to 0.7 nm with heating at 300 °C (Fig. 3.6c). The 0.7 nm basal spacing of kaolinite and halloysite disappear with heating at 550 °C because these clays are converted to a non-crystalline phase. The CEC of halloysite ranges up to 40 $cmol_c\ kg^{-1}$, and some halloysites show high selectivity for K^+ and NH_4^+.

Kaolin minerals were described in detail by Dixon (1989). Recently, halloysite was reviewed by Joussein et al. (2005) and by Churchman et al. (2016).

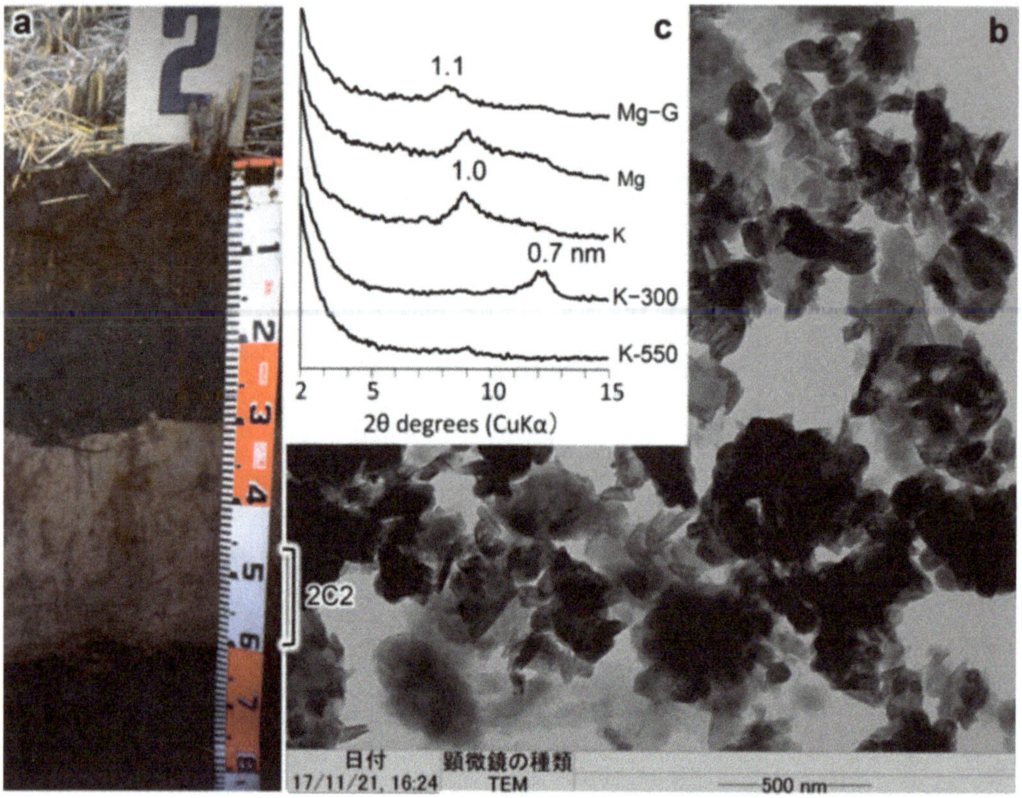

Fig. 3.6 Halloysite in partially weathered volcanic ash horizons. (**a**) A soil profile. A kaolin-rich clay fraction was prepared from the 2C2 horizon. (**b**) TEM image, (**c**) XRD patterns with five treatments (Mg saturation with glyceration, Mg saturation, K saturation, heating at 300 °C after K saturation, and heating at 550 °C after K saturation) of the dithionite–citrate–bicarbonate treated and oriented clay fraction

3.2.3.2 2:1 Type Minerals

Major 2:1 type minerals include smectite, vermiculite, micaceous minerals, chlorite, etc. (Table 3.1, Fig. 3.3). Among these minerals, smectite and vermiculite show high CEC, ranging between 90 and 150 $cmol_c\ kg^{-1}$. Soils having these clay minerals and neutral-range pH are fertile because the minerals can carry a large amount of nutrient cations due to their high CEC.

Smectite is a group name including three members, montmorillonite, beidellite, and nontronite. The differences of the members are the major site of isomorphous substitution and the ions at the sites of substitution. Montmorillonite has major isomorphous substitution at the octahedral sites, whereas beidellite has it at the tetrahedral sites. Montmorillonite can be distinguished from other smectites by using the Greene-Kelly test (Greene-Kelly 1955; Mala and Douglas 1987). If beidellite has a large amount of iron at the octahedral sites, it is called nontronite (Table 3.1).

Smectite looks like thin sheets. Figure 3.7a, a TEM image of a reference smectite, suggests that many thin smectite sheets are randomly overlapped. Figure 3.7b, an

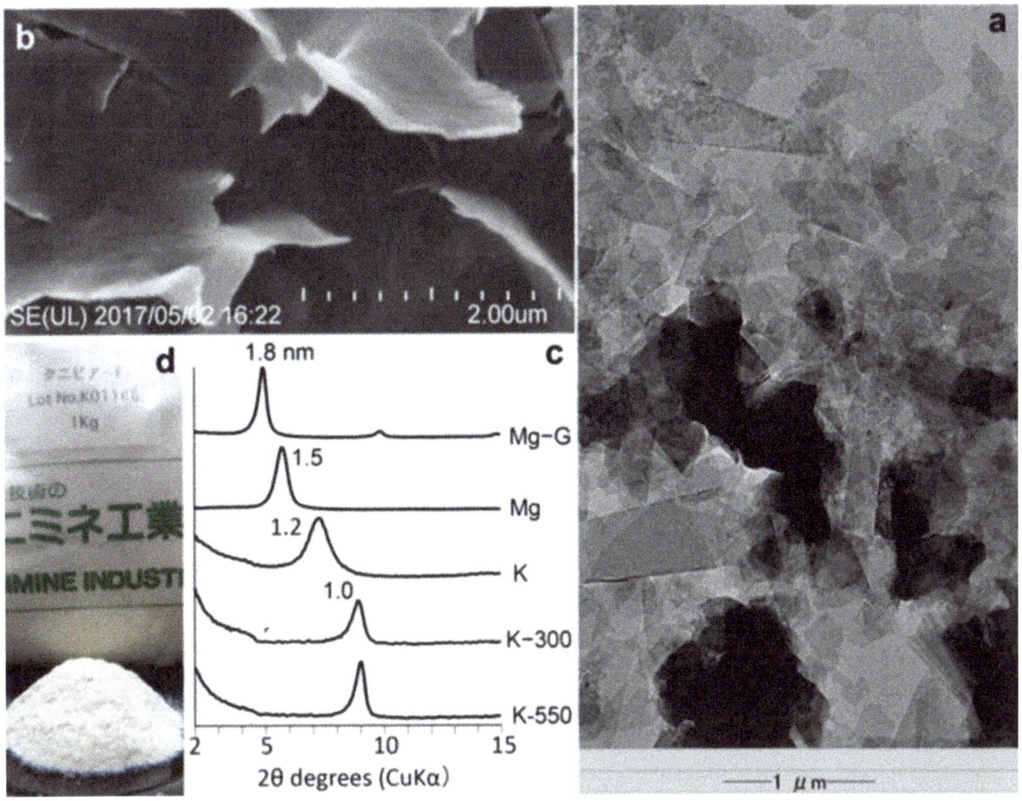

Fig. 3.7 Reference montmorillonite. (**a**) TEM image, (**b**) SEM image, (**c**) XRD patterns of the oriented clay fraction of Kunipia F (**d**) with five treatments (Mg saturation with glyceration, Mg saturation, K saturation, heating at 300 °C after K saturation, and heating at 550 °C after K saturation). Kunipia F, is natural montmorillonite produced at Tsukinuno, Yamagata Prefecture, Japan

SEM image showing the surface of the overlapped sheets, suggests that some thin sheets are turned up.

The basal spacing of smectite expands to accommodate water or organic compounds in the interlayer sites. In other words, the smectite swells. When the exchangeable cation is Mg^{2+}, the basal spacing is 1.4–1.5 nm (Fig. 3.7c), which is due to the six-coordination of water molecules around a divalent cation. The arrangement of six water molecules corresponds to two layers of water molecules at the interlayer site. The basal spacing of Mg^{2+}-saturated smectite (1.5 nm) expands to 1.8 nm with the treatment by glycerin. This property is used to identify smectite in soil.

Similar electron micrographs are obtained from smectite-rich soil (Fig. 3.8). The TEM image (Fig. 3.7a) shows overlapped thin sheets, and the SEM image suggests turned up thin sheets (Fig. 3.7b). The XRD peaks for the basal spacing are broader than those in Fig. 3.7c. A small diffraction peak at 0.7 nm (Fig. 3.8c) is probably from kaolinite, a minor component of this clay fraction.

Montmorillonite shows higher swelling than the other two members of the smectite group (Greene-Kelly 1955; Egashira and Ohtsubo 1983). Both smectites,

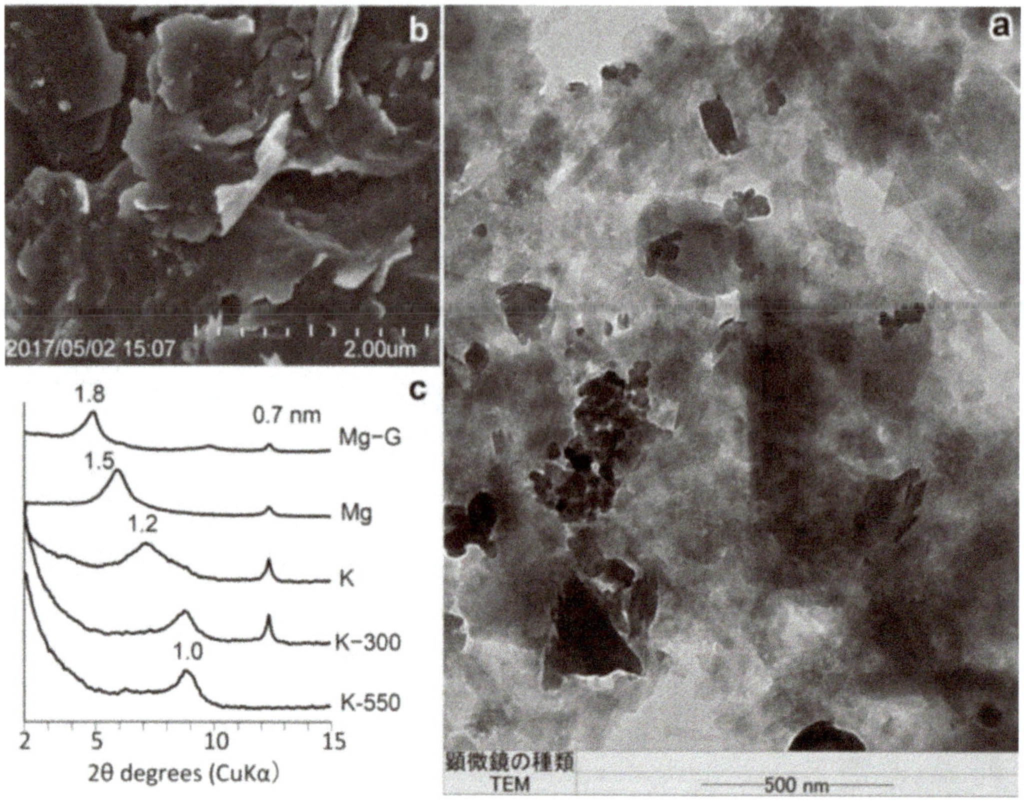

Fig. 3.8 Montmorillonite from subsoil of a paddy field. (**a**) TEM image, (**b**) SEM image, (**c**) XRD patterns with five treatments (Mg saturation with glyceration, Mg saturation, K saturation, heating at 300 °C after K saturation, and heating at 550 °C after K saturation) of the dithionite–citrate–bicarbonate treated, and oriented clay fraction. The clay fraction was obtained from the 25–40 cm horizon (Bg2) of the pedon shown in Fig. 5.22b

Figs. 3.7 and 3.8, are montmorillonite because the basal spacing is about 0.98 nm by the Greene-Kelly test (Li^+ saturation, heating at 250 °C, and glyceration). Montmorillonite has a greater amount of isomorphous substitution at the octahedral sites than the other smectites. The layer charge of montmorillonite is diminished by Li^+, which can possibly penetrate into the cation deficient octahedral sites due to its small ionic size. As a result, the swelling property of montmorillonite is diminished. When it is saturated with K^+ and heated at 300 °C followed by glyceration, the basal spacing expands to 1.8 nm, which is also a property of montmorillonite (Malla and Douglas 1987).

Soils with abundant smectite show an intensive shrink–swell cycle under semi-dry climate, in which a dry season alternates with a wet season. In the dry season, wide cracks form from the soil surface to a depth of several tens of cm or deeper. The cracks may develop horizontally or diagonally, not only vertically. When the rainy season begins, water enters the lower horizons of the soil along the cracks and swelling begins at all sites having moisture. The swelling force at the lower horizon may lift the upper soil, and the soil horizon may become undulated. Uneven swelling at the lower horizon makes a sliding surface in the soil, a slickenside. These are the properties of Vertisols, which occupy about 3% of the world's soil and are distributed mainly in a portion of the Mexico Gulf region, a part of India, the upper Nile flow region, and southeastern Australia. Without alternating wet and dry seasons, Vertisols do not form even if smectite is abundant in the soil.

The characteristics of vermiculite are high CEC, dominance of the trioctahedral type, and a wide range of particle size from sand to clay. The basal spacing of Mg^{2+}-saturated vermiculite is 1.4 nm. The spacing decreases to 1.0 nm with K^+ saturation, and it does not expand with glyceration. This property is used to identify vermiculite in soil from XRD patterns. The trioctahedral vermiculite can form from trioctahedral micas in a tephra-deived soil (Nanzyo et al. 2001).

Figure 3.9 shows the TEM image, SEM image, and XRD patterns of the clay fraction prepared from the Bw horizon of the soil profile shown in Fig. 6.24. The TEM image (Fig. 3.9a) suggests that this clay fraction has mixed clay minerals. The major ones are vermiculite (platy) and halloysite (curved tubular).

One of the minimal swelling members of the 2:1 minerals is micaceous minerals (Fig. 3.3) including illite (Table 3.1). The basal spacing of micaceous minerals is 1.0 nm, and the basal spacing hardly changes with heating at 300 or 550 °C. Negative charges of the aluminosilicate layer are neutralized with K^+, which is fixed in the interlayer site. The interlayer K^+ is exchanged little with other cations.

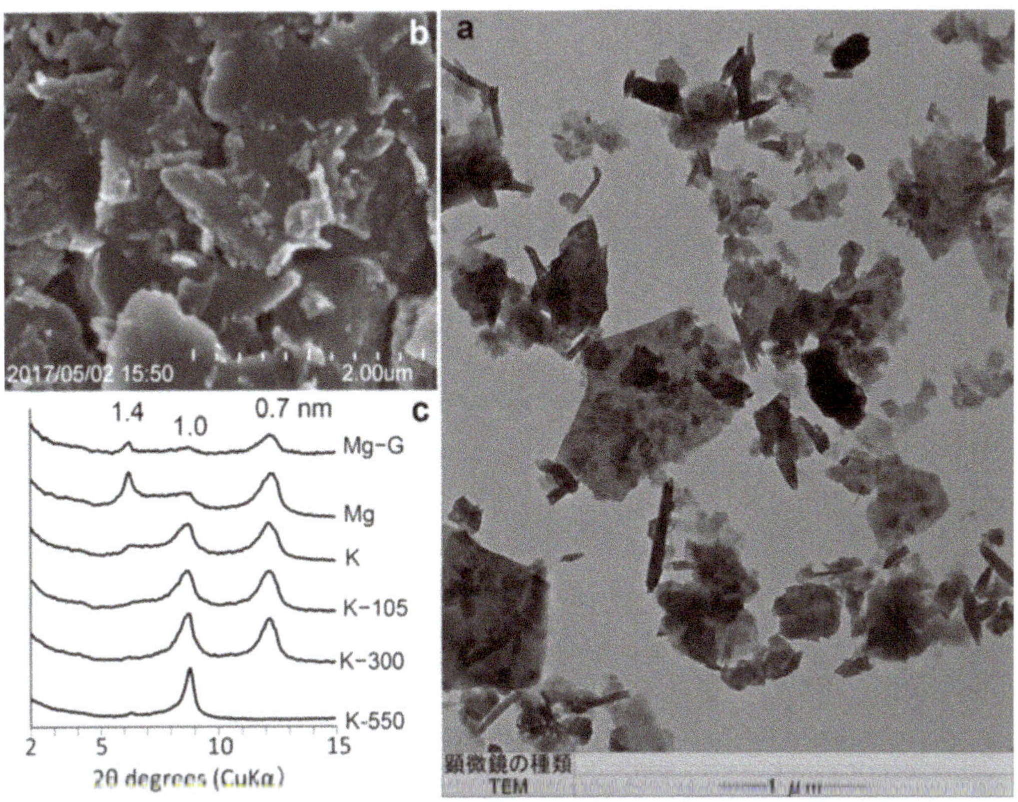

Fig. 3.9 The Clay fraction obtained from 7–22 cm horizon (Bw) of the pedon shown in Fig. 6.24. (**a**) TEM image, (**b**) SEM image, (**c**) XRD patterns with six treatments (Mg saturation with glyceration, Mg saturation, K saturation, and heating at 105, 300, and 550 °C after K saturation, respectively) of the oriented vermiculite-containing clay fraction

The particle size of micaceous minerals ranges widely between sand Figs. 2.10 and 2.11 and clay. Micaceous minerals are also formed by diagenesis, not only from the weathering of igneous rocks.

The final member of the 2:1 minerals is chlorite (Fig. 3.3). Chlorite has an inorganic polymer at the interlayer site. For example, in Mg chlorite (Table 3.1), negative charges of the aluminosilicate layer accommodate a positively charged brucite layer in which Mg^{2+} is partly replaced with Al^{3+}. Chlorite can be regarded as a 2:1:1 type mineral, in which the last layer is, for example, the brucite layer of Mg chlorite.

As an example of chlorite, a clay fraction prepared from partially weathered chlorite schist (metamorphic rock) is shown in Figs. 3.10a, b. As shown in the SEM image (Fig. 3.10b) of the clay fraction, chlorite is also thin plates, but these plates are thicker than those of smectite. The basal spacing of chlorite is 1.4 nm (Fig. 3.3). Although this basal spacing is not affected by Mg^{2+} or K^+ saturation, the peak intensity of the 1.4 nm reflection increases with heating at 550 °C (Fig. 3.10c).

In acid soils, types intermediate between vermiculite (or smectite) and chlorite are often found. These are chloritized vermiculite and chloritized smectite. These

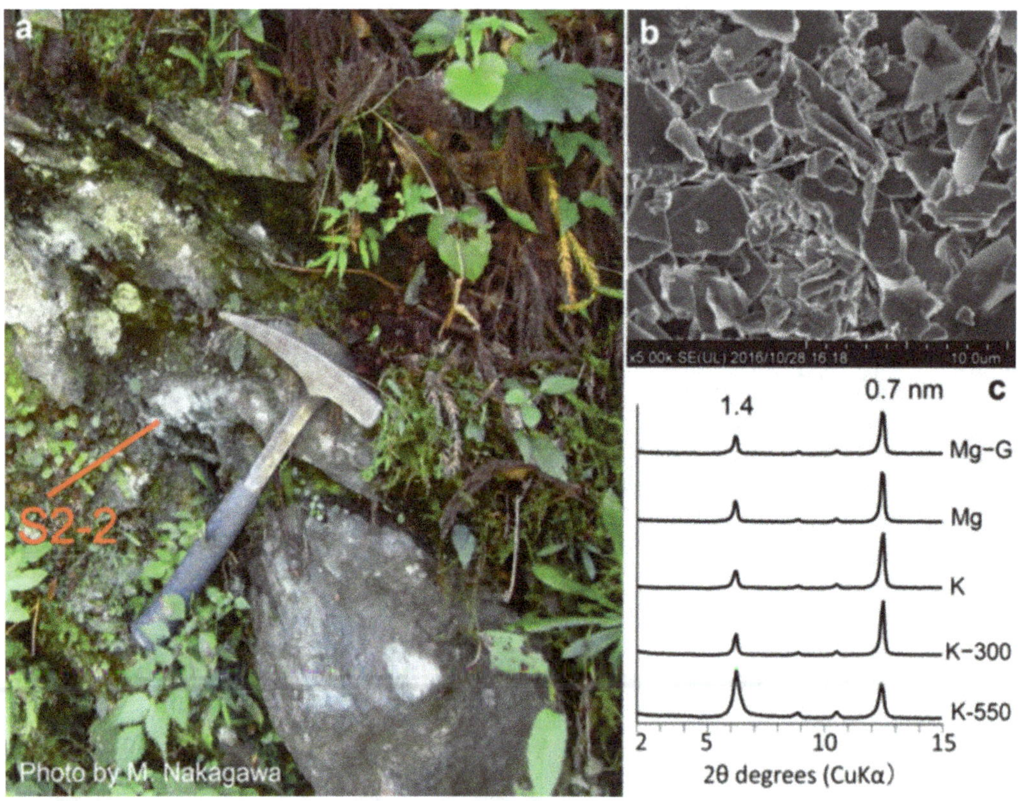

Fig. 3.10 (**a**) Partially weathered chlorite schist, (**b**) SEM image of the chlorite-rich clay fraction prepared from the position S2–2 of (**a**), (**c**) XRD patterns with five treatments (Mg saturation with glyceration, Mg saturation, K saturation, heating at 300 °C after K saturation, and heating at 550 °C after K saturation) of the oriented clay fraction. The chlorite schist was sampled near the boundary between the chlorite and gabbro zones along the Asemi-gawa River (Higashino 1990; Taguchi and Enami 2014) by M. Nakagawa

minerals appear to form through polymerization of exchangeable Al in an acid soil, but the polymerization of Al is not complete as in a gibbsite sheet. The position of the XRD peak of the basal spacing appears broadly between 1.4 and 1.0 nm after heating at 300 °C. Several researchers have used citrate, fluoride, etc., to remove the polymerized Al interlayering (Barnhisel 1977).

Acid soils, soils with pH(H_2O) roughly less than 5.5, have exchangeable or KCl-extractable Al and cause Al-overload disorders in sensitive plants. Liming is effective for removing this exchangeable Al. Further, application of other plant nutrients, such as Mg, K, P, etc., is necessary to improve crop production in acid soils.

3.2.4 Dioctahedral and Trioctahedral Type

The difference in the octahedral layers between dioctahedral and trioctahedral types appears to affect the weathering resistance of 2:1 type aluminosilicates. By comparing the muscovite and the weathered biotite in Fig. 2.9, biotite appears to weather earlier than muscovite in soil. The octahedral layer of biotite is the trioctahedral type, and that of muscovite is the dioctahedral type.

Figure 3.11 schematically shows the differences between trioctahedral and dioctahedral micas. Potassium ion locates at the center of two six-membered Si tetrahedron rings and is sandwiched by both upper and lower phyllosilicate layers, as shown in Fig. 3.11a, b. In the trioctahedral type, through observation of the lower aluminosilicate shown in Fig. 3.11a from the plane shown in Fig. 3.11c, the six-membered Si tetrahedron ring and a part of the lower trioctahedral layer appear (Fig. 3.11d). At the center of the Si tetrahedron ring, there is a front OH. The direction of the proton of the front OH is perpendicular to the phyllosilicate plane because the front OH is surrounded by three Al ions. Due to the difference in electronegativity between O and H, the H is positively charged slightly. Hence, a repulsive force occurs between K^+ and $H^{\delta+}$. On the other hand, in the dioctahedral type, one-third of the octahedral sites are vacant, and the proton of the front OH can be directed to the vacant octahedral site (Fig. 3.11e). Returning to Fig. 3.11a, b, the repulsion between K^+ and $H^{\delta+}$ is weaker in the dioctahedral type than in the trioctahedral type. This is the possible reason why dioctahedral mica is more stable than trioctahedral mica in soil.

The position of the powder XRD peak of the (060) plane can be used to distinguish the trioctahedral type from the dioctahedral type. The positions of the XRD peak of these two types are 1.53 and 1.50 nm, respectively. If a clay sample is a mixture of the two types, it is difficult to make this distinction. When the mixture is with kaolin minerals, peaks from the kaolin minerals can be removed by heating at 550 °C. If the composition of the remaining mineral is simple enough, distinction between these two types using the position of the (060) peak may be possible (Nanzyo et al. 2001).

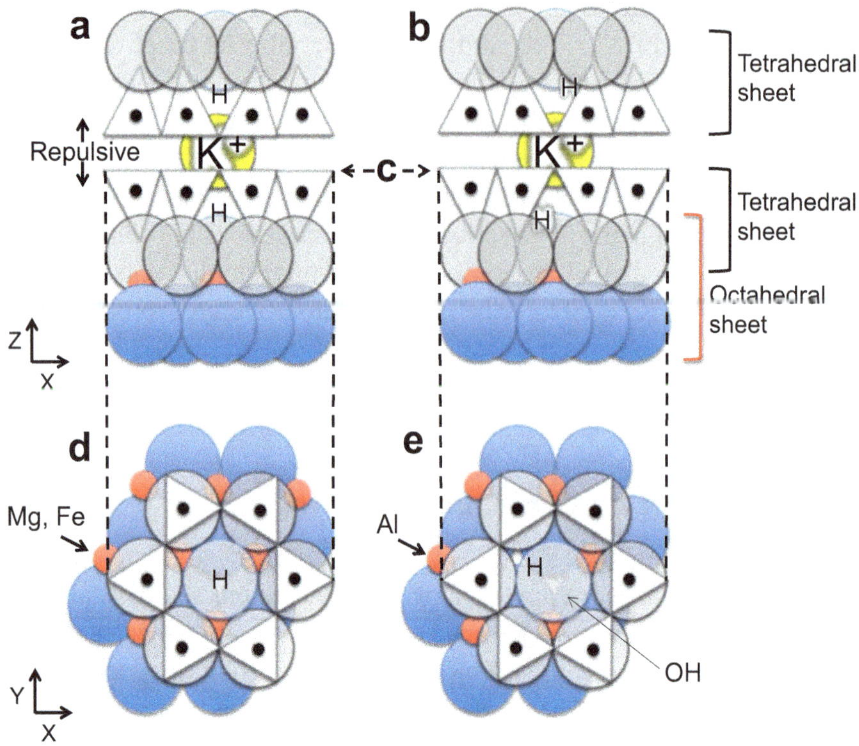

Fig. 3.11 Schematic showing the differences between biotite (left) and muscovite (right). (**a** and **b**) Potassium ion sandwiched by both upper and lower phyllosilicate layers, (**c**) the plane to show the lower aluminosilicate (**d** and **e**)

3.3 Oxides, Hydroxides, and Others

Secondary minerals in soil other than phyllosilicates include oxides, hydroxides, carbonates, sulfates, phosphates, halides, etc., and these are summarized in Table 3.2. Some of these are reaction products between soil inorganic constituents and fertilizers added to farmland. In this section, gibbsite and manganese oxides are considered. The minerals in Table 3.2 include minerals introduced in subsequent chapters.

Gibbsite

Gibbsite is an aluminum hydroxide [$Al(OH)_3$] mineral occurring in highly weathered soils such as Oxisols (Hsu 1989). The gibbsite sheet was shown in Fig. 3.2b. Gibbsite sheets stack in the Z direction with hydrogen bonding.

Gibbsite also occurs in Andisols (Wada and Aomine 1966). Figure 3.12a, b shows an Andisol having a gibbsite-containing horizon at the depth of 66–85 cm from the surface. Gibbsite exists in the 0.2–0.05, 0.05–0.02, and 0.02–0.005 mm particle size fractions as shown by the XRD peaks at the d-spacing of 0.485 and 0.437 nm.

In the 0.05–0.02 mm particle size fraction, gibbsite particles were found using element mapping; they appear as Al-rich particles. In Fig. 3.13, there are two

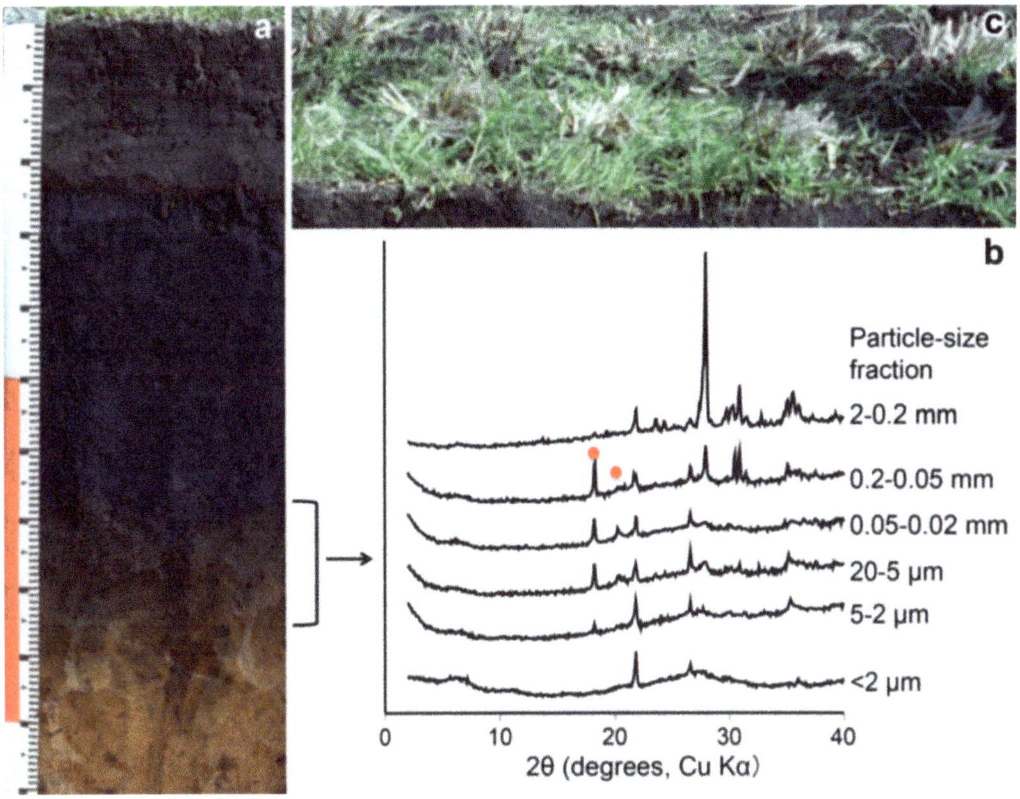

Fig. 3.12 Preparation of a gibbsite sample. (**a**) An Andisol profile having a gibbsite-containing horizon at the depth of 66–85 cm, (**b**) XRD patterns of six particle size fractions of the gibbsite-containing soil horizon, (**c**) the present land use (paddy field, after harvest). Red closed circles indicate the highest (d = 0.485 nm) and second highest (d = 0.437 nm) XRD peaks of gibbsite

gibbsite particles marked by the letters (a) and (b). As shown by the EDX spectra (Fig. 3.13a, b), the cations of these particles are mostly Al. Other particles in Fig. 3.13 are aggregates of allophanic non-crystalline materials (most abundant, Fig. 3.13c) and non-colored volcanic glass (Fig. 3.13d). Because allophanic materials and volcanic glass show no sharp XRD peaks, the 0.485 nm peak from gibbsite appears as the highest peak in the 0.05–0.02 mm fraction (Fig. 3.12b). It is deducible that with intensive leaching of Si, Ca, Na, etc., from volcanic glass, Al-rich particles such as allophanic materials and gibbsite formed.

Gibbsite particles appear to have hexagonal platy characteristics (Hsu 1989). Figure 3.14 is the magnified SEM image of the particle (b) shown in Fig. 3.13. The lower part of the particle shown in Fig. 3.14 has hexagonal platy characteristics.

Manganese Oxides and Hydroxides

Manganese is one of the major elements in soil and also one of the essential elements for plants, animals, and other organisms. Macroscopically, manganese oxides and hydroxides appear in soil as nodules. Small patchy concentrated manganese is sometimes found in the subsoil of irrigation water paddy field. However, it is not easy to separate manganese minerals unless they are not concentrated as nodules or concretions.

Fig. 3.13 SEM image of the 0.05–0.02 mm fraction of the gibbsite-containing soil horizon at the depth of 66–85 cm of the pedon shown in Fig. 3.12a. EDX spectra (**a**), (**b**), (**c**), and (**d**) were obtained from the dashed areas (**a**), (**b**), (**c**), and (**d**) of the different particles, respectively

Fig. 3.14 Magnified SEM image of the gibbsite particle labeled as Fig. 3.13b

From among the manganese minerals reviewed by McKenzie (1989), the examples of birnesite and lithiophorite are introduced in this section. Birnesite is a group of manganese oxides including δ-MnO_2, manganous manganite, and others. According to McKenzie (1989), the birnesite group is characterized by a 0.7 nm XRD basal reflection and its higher order. The chemical formula of lithiophorite is $(Al,Li)MnO_2(OH)_2$. Analyses of lithiophorite and birnesite show high Mn_3O_4 concentration (McKenzie 1989).

Manganese nodules were collected from the soil surface of Yomitan-son, Okinawa Prefecture, Japan (Fig. 3.15a, b) by hand-picking. The Mn nodules are dark-colored balls with a diameter of about 1 cm (Fig. 3.15b). Tokashiki et al. (2003) used a sequential selective dissolution method to separate the birnesite, lithiophorite, and also goethite in soil manganese nodules. X-ray diffraction patterns were used to confirm the presence of these minerals. To obtain visual information for these manganese minerals, polished sections of the Mn nodules (Fig. 3.15c) were prepared. Following the sequential selective dissolution method of Tokashiki et al. (2003), a small diffraction peak at 0.72 nm (Fig. 3.15d), which appeared after 5 M

Fig. 3.15 (**a**) Soil profile, (**b**) manganese nodules collected from the surface of soil (**a**), (**c**) polished section of the manganese nodules, (**d**, **e**, **f**, and **g**) XRD patterns with four sequential treatments of the ground manganese nodules (**d**: after 5 M NaOH treatment, **e**: after hydroxyl amine hydrochloride treatment (HAHC) at 25 °C, **f**: after HAHC at 60 °C, **g**: after dithionite–citrate–bicarbonate treatment). The XRD patterns were obtained by air-drying a sample suspension on a slide glass

NaOH treatment of powdered Mn nodule and disappeared after hydroxylamine hydrochloride treatment (HAHC) at 25 °C, suggests birnesite. The intensity of diffraction peaks at 0.947 and 0.472 nm increased with the HAHC treatment at 25 °C and disappeared with the HAHC treatment at 60 °C, suggesting lithiophorite. The intensity of diffraction peaks at 0.418, 0.268, 0.258, and 0.245 nm, which disappeared with dithionite–citrate–bicarbonate (DCB), suggests goethite. The diffraction peaks that remained after the DCB treatment, 1.00, 0.501, and 0.334 nm, correspond to those of a micaceous mineral. All of these results are close to those reported by Tokashiki et al. (2003).

In the optical micrograph (Fig. 3.16a) of a polished section, concentric color distribution patterns of brown and dark - gray can be seen. By comparing the optical photograph (Fig. 3.16a) with the Fe element map (Fig. 3.16b), the brown color can be approximately correlated with iron concentration. Among these concentric color and iron distribution patterns, Mn concentration is higher in the dark - gray parts (Fig. 3.16d) than in the brown iron-rich parts (Fig. 3.16c, e). In the magnified Mn element map (Fig. 3.16f) of the dashed square in Fig. 3.16a, Mn-concentrated spots occur in the lower-right dashed square.

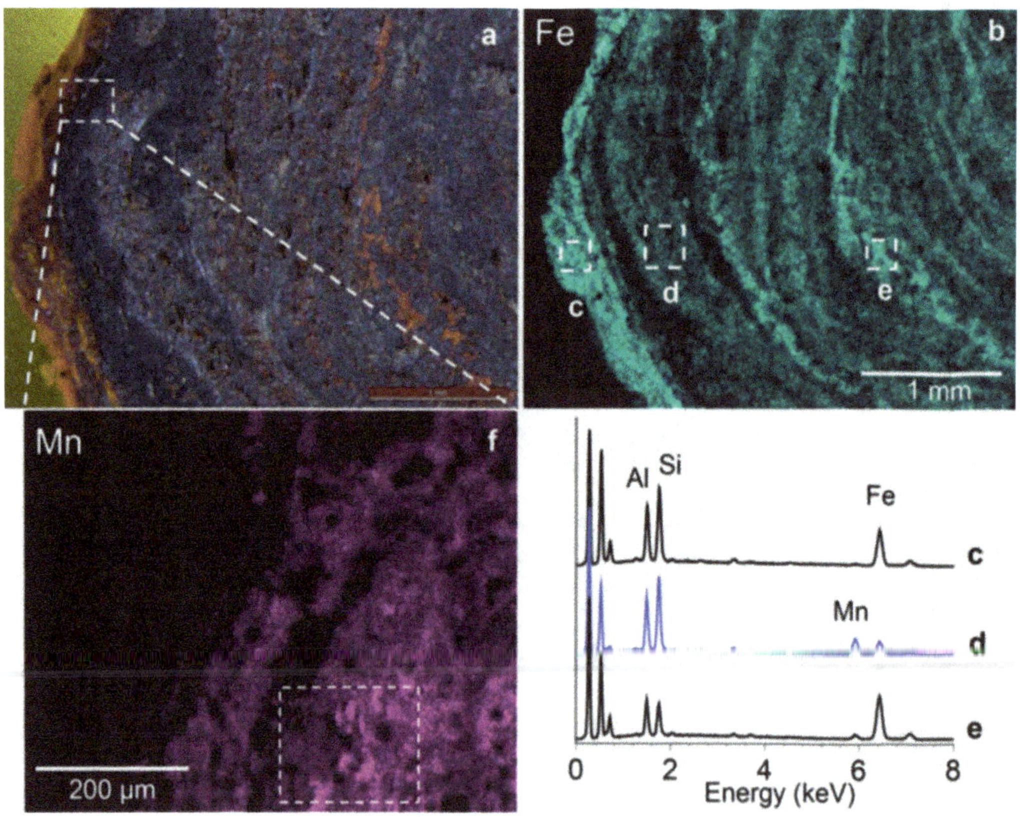

Fig. 3.16 EDX analyses of a manganese nodule. (**a**) Optical photograph of polished section showing a magnified view of the left-hand side of the upper left nodule in Fig. 3.15c, (**b**) an element map showing Fe distribution of (**a**), (**c, d,** and **e**) EDX spectra obtained from the dashed squares (**c**), (**d**) and (**e**) in (**b**), (**f**) a magnified element map showing Mn distribution of the dashed square in (**a**)

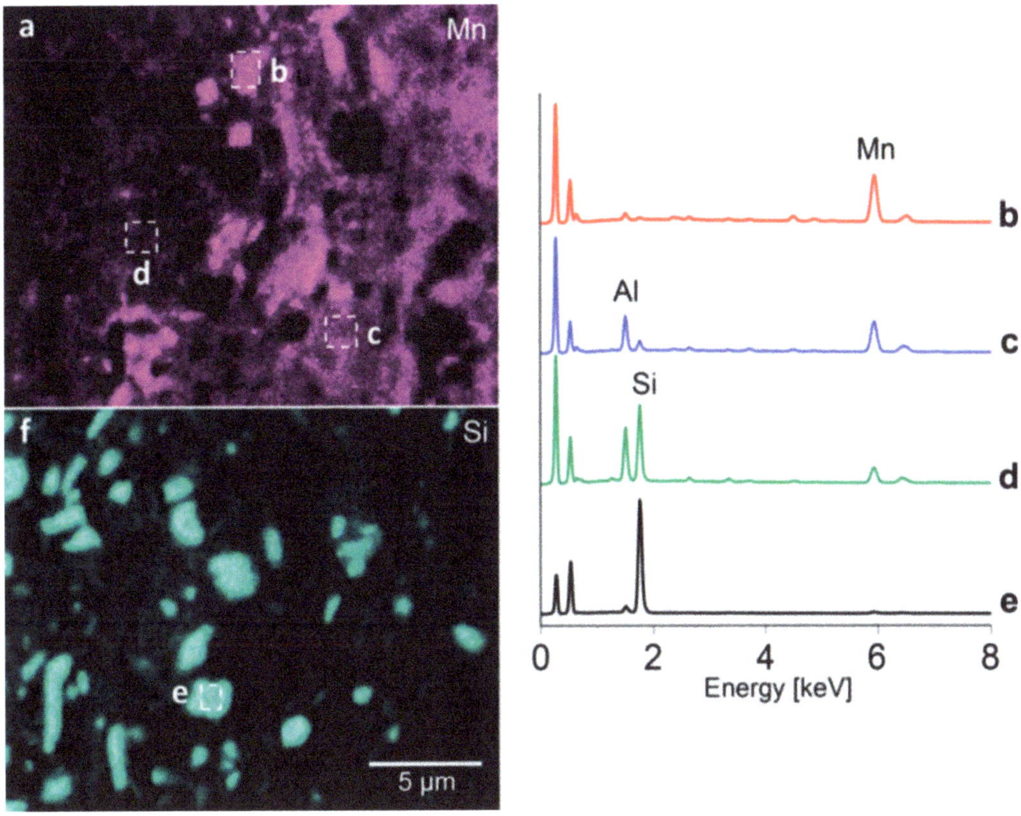

Fig. 3.17 EDX analyses with higher magnification. (**a**) magnified element map of Mn of the dashed square in Fig. 3.16f, (**b**, **c**, and **d**) EDX spectra obtained from the dashed squares (**b**), (**c**), and (**d**) in (**a**), (**e**) EDX spectrum obtained from the dashed square (**e**) in the Si element map (**f**) of the same area as in (**a**)

Further magnification of the dashed square in Fig. 3.16f gave Fig. 3.17a, in which different Mn concentration areas can be found. In the highest Mn concentration areas, for example the dashed square (b), the concentration of Mn is much higher than the concentration of other elements, as shown in the EDX spectrum shown in Fig. 3.17b. In the second-highest Mn concentration areas, for example the dashed area (c), Mn and Al concentrations are high, as shown in the EDX spectrum shown in Fig. 3.17c. In the third-highest Mn concentration areas, for example the dashed area (d), Si and Al concentrations are much higher than Mn concentration, as shown by the EDX spectrum shown in Fig. 3.17d. In the lowest Mn concentration areas, Si concentration is high, as shown by the Si element map (Fig. 3.17f) and the EDX spectrum shown in Fig. 3.17e. Considering the analyses of manganese minerals (Mn_3O_4: 83.2 and 73.2%; Al_2O_3: 1.0 and 11.5% for birrnessite and lithiophorite, respectively (McKenzie 1989)) and the results from the sequential selective dissolution analyses (Fig. 3.15), the major component of the dashed areas shown in Fig. 3.17b, c can be related to birnesite and lithiophorite, respectively. The major components of the areas of the EDX spectra shown in Fig. 3.17d, e can be related to a dioctahedral micaceous mineral (Fig. 3.15g) and fine silica particles, respectively.

As suggested by the small peaks in EDX spectra (b), (c), (d), and (e), those dashed areas in Fig. 3.17a, f may be a mixture with other minor components.

Further readings for this chapter are Dixon and Weed (1989) and Dixon and Schulze (2002).

References

Barnhisel RI (1977) Chlorites and hydroxyl interlayered vermiculite and smectite. In: Dixon JB, Weed SB, Kttrick JA, Milford MH, White JL (eds) Minerals in soil environment. SSSA, Madison

Churchman GJ, Lowe DJ (2012) Alteration, formation, and occurrence of minerals in soils. In: Huang PM, Li Y, Sumner ME (eds) Handbook of soil sciences, properties and processes, vol 20, 2nd edn. CRC Press/Taylor & Francis Group, Boca Raton/London/New York, pp 1–72

Churchman GJ, Pasbakhsh P, Hillier S (2016) The rise and rise of halloysite. Clay Miner 51:303–308

Deer WA, Howie RA, Zussman J (2013) An introduction to the rock-forming minerals, 3rd edn. Mineralogical society, London

Dixon JB (1989) Kaolin and serpentine group minerals. In: Dixon JB, Weed SB (eds) (Co-eds) minerals in soil environments, 2nd edn. Soil Science Society of America, Madison, pp 467–525

Dixon JB, Schulze DG (eds) (2002) Soil mineralogy with environmental applications. SSSA book series, no. 7. SSSA, Madison

Dixon JB, Weed SB (eds) (1989) Minerals in soil environment. SSSA book series no. 1. SSSA, Madison

Egashira K, Ohtsubo M (1983) Swelling and mineralogy of smectites in paddy soils derived from marine alluvium, Japan. Geoderma 29:119–127

Greene-Kelly R (1955) Dehydration of montmorillonite minerals. Mineral Mag 30:604–615

Harris W, White GN (2008) X-ray diffraction techniques for soil mineral Identification. In: Ulery A, Vepraskas M, Wilding L (eds) Methods of soil analysis. Part 5. Mineralogical methods. SSSA book series, no. 5. SSSA, Madison, pp 81–115

Higashino T (1990) The higher grade metamorphic zonation of the Sambagawa metamorphic belt in central Shikoku. Japan J Metam Geol 8:413–423

Hsu PH (1989) Aluminum hydroxides and oxyhydroxides. In: Dixon JB, Weed SB (eds) (Co-eds) Minerals in soil environments, 2nd edn. Soil Science Society of America, Madison, pp 331–378

Joussein E, Petit S, Churchman J, Theng B, Righi D, Delvaux B (2005) Halloysite clay minerals – a review. Clay Miner 40:383–426

Kampf N, Scheinorst AC, Schulze DG (2012) Oxide minerals in soils. In: Huang PM, Li Y, Sumner ME (eds) Handbook of soil sciences, properties and processes, vol 22, 2nd edn. CRC Press/Taylor & Francis Group, Boca Raton/London/New York, pp 1–34

Kodama H (2012) Phyllosilicates. In: Huang PM, Li Y, Sumner ME (eds) Handbook of soil sciences, properties and processes, vol 21, 2nd edn. CRC Press, Taylor & Francis group, Boca Raton/London/New York, pp 1–49

Malla PB, Douglas LA (1987) Problems in identification of montmorillonite and beidellite. Clay Clay Miner 35:232–236

McKenzie RM (1989) Manganese oxides and hydroxides. In: Dixon JB, Weed SB (eds) (Co-eds) Minerals in soil environments, 2nd edn. Soil Science Society of America, Madison, pp 439–465

Molloy MW, Kerr PF (1961) Diffractometer patterns of A.P.I. reference clay minerals. Am Mineral 46:583–605

Nanzyo M, Tsuzuki H, Otuska H, Yamasaki S (2001) Origin of clay-size vermiculite in sandy volcanic ash soils derived from modern Pinatubo lahar deposits in Central Luzon, Philippines. Clay Sci 11:381–390

Soil Survey Staff (1999) Soil taxonomy, a basic system of soil classification of making and interpreting soils surveys, USDA-NRCS, agriculture handbook no. 436, U.S. Government Printing Office, Washington DC

Sudo T, Shimoda S (1978) Clays and clay minerals of Japan. Developments in sedimentology 26, Elsevier, Amsterdam coprinted by Kodansha LTD., Tokyo, pp 1–326

Taguchi T, Enami M (2014) Compositional zoning and inclusion of garnet in Sanbagawa metapelites from the Asemi-gawa route, central Shikoku. Japan J Mineral Petrol Sci 109:1–12

Tokashiki Y, Hentona T, Shimo M, Arachchi LPV (2003) Improvement of the successive selective dissolution procedure for the separation of birnessite, lithiophorite, and goethite in soil manganese nodules. Soil Sci Soc Am J 67:837–843

Wada K, Aomine S (1966) Occurrence of gibbsite in weathering of volcanic materials at Kuroishbaru, Kumamoto. Soil Sci Plant Nutr 12:151–157

Chapter 4
Non-crystalline Inorganic Constituents of Soil

Abstract Non-crystalline inorganic constituents of soil, such as volcanic glasses, phytoliths, laminar opaline silica, allophane, and imogolite are introduced using optical and electron microscope images and energy dispersive X-ray (EDX) analysis. The Al-humus complex and Al-rich Sclerotia grains are also introduced. The volcanic glasses are formed from magma and can be categorized as primary. All of these non-crystalline inorganic constituents are found in volcanic ash soils. Among these, phytoliths can be found under vegetation in many other soils than volcanic ash soils. Formation of allophanic materials from fresh pumice is shown stepwise using polished sections to demonstrate microscopic distribution of elements and inorganic constituents. Allophane and imogolite are rich in Al whereas their parent material, volcanic ash, is silica-rich. Changes in morphological property and element concentration of volcanic ash or volcanic glass during the formation of these secondary non-crystalline constituents are discussed.

4.1 Introduction

The non-crystalline constituents present in soil depend strongly on soil environmental conditions. The non-crystalline silicate and silica constituents in soil are volcanic glass, allophane, imogolite, laminar opaline silica, and phytoliths (Table 4.1). All these are present in volcanic ash soils. Among these materials, allophane and imogolite are also found in spodic horizons, such as Bs and Bhs horizons, and they confer unique properties upon Andisols and Spodosols. Volcanic ash soils are found in volcanic areas worldwide. Spodosols are found mainly in the cold regions of the world, and can also be formed in volcanic ash deposits. Phytoliths form in plant cells and occur commonly in many A and buried A horizon soils.

The Al-humus complex is also covered in this chapter despite its partially organic structure because it is one of the major constituents of active Al. Active Al is important to the formation of Andisols and Spodosols, classification of these soils, and the phosphate sorption reaction.

Other non-crystalline materials which may be present in soils include non-crystalline phases of iron sulfide, iron phosphate, and aluminum phosphate.

M. Nanzyo, H. Kanno, *Inorganic Constituents in Soil*,
https://doi.org/10.1007/978-981-13-1214-4_4

Table 4.1 Non-crystalline silicate and silica constituents in soil

Non-crystalline constituents	Chemical formula
Volcanic glass	
Allophane	$1\text{-}2SiO_2{\cdot}Al_2O_3{\cdot}nH_2O$
Imogolite	$(OH)_2Si_2O_6Al_4(OH)_6$
Laminar opaline silica	$SiO_2{\cdot}nH_2O$
Phytolith	$SiO_2{\cdot}nH_2O$

Non-crystalline iron sulfide and iron phosphate were covered in Chap. 5, and non-crystalline aluminum phosphate were covered in Sect. 6.4. The term "volcanic ash soils" encompasses all soils derived from volcanic ash, whereas Andisols are soils defined by Soil Taxonomy of United States Department of Agriculutre (USDA).

4.2 Volcanic Glass

Volcanic ash, in this monograph, refers collectively to volcanic ejecta or tephra, including pyroclastic fall and flow materials such as volcanic ash, cinders, lapilli, scoria, and pumice (Dahlgren et al. 1993). Volcanic glass is a non-crystalline silica-alumina material and is a major constituent of the volcanic ash ejected from volcanoes (Yamada and Shoji 1975; Heiken and Wohletz 1985). The morphological types of volcanic glasses include sponge-like, bubble-wall type (curved platy), fibrous, and berry-like (Shoji et al. 1993). The diameter of sponge-like and berry-like volcanic glasses can exceed 2 mm. The diameter of volcanic glass ranges below 2 μm. The color of volcanic glasses is related to the rock type of volcanic ash. Sponge-like, bubble-wall, or fibrous glasses are non-colored. Coloured volcanic glass is mostly berry-like with crystallites.

4.2.1 Chemical Composition of Volcanic Glasses

Shoji et al. (1975) classified volcanic ashes into five rock types based on the total SiO_2 content. The five rock types are rhyolite, dacite, andesite, basaltic andesite, and basalt. Their total SiO_2 content ranges are 100–70, 70–62, 62–58, 58–53.5, and 53.5–45%, respectively. Reclassification of 26 tephra reported by Shoji et al. (1975) using the updated classification of Le Maitre (2002), known as the total alkali silica classification (TAS diagram), yielded the same results for 22 tephra. Two of the remaining tephra were close to the boundary between andesite and dacite, and the other two lay between dacite and rhyolite, because the $Na_2O + K_2O$ contents of all 26 tephra were below the boundaries between the four rock types (basalt, basaltic andesite, andesite, and dacite) and the corresponding trachy-types of the TAS diagram.

Volcanic ash is sorted by its particle size and the specific gravity of its constituent mineral particles during transportation in air. In the case of Tarmae-a (Ta-a) tephra (AD1739), Hokkaido, Japan, the heavy mineral content decreased with distance from source volcano. As the heavy mineral content decreased, the Fe and Mg content of Ta-a tephra also decreased while the Si content increased correspondingly (Mizuno et al. 2008).

The color of volcanic glasses is related to the rock type of the tephra. Volcanic ash of rhyolite, dacite, and andesite rock types is dominated by non-colored volcanic glasses including a slightly purplish one, whereas basalt and basaltic andesite rock type is dominated by colored volcanic glasses (Shoji 1986). The chemical composition of volcanic ash is also related to rock types. The Al_2O_3, FeO, MgO, and CaO content of 26 volcanic ash samples significantly, and Na_2O, K_2O and TiO_2 also correlated with SiO_2 content (Shoji et al. 1975). In nine non-colored volcanic glass samples, SiO_2 content ranged between 74.24 and 77.89%, and Al_2O_3 content ranged between 12.65 and 14.71%. The variation in the concentrations of SiO_2 and Al_2O_3 of the non-colored volcanic glasses are minor compared to that for the total SiO_2 and Al_2O_3 concentrations of the volcanic ashes. In contrast, in glasses present in basaltic andesite and basalt volcanic ashes, the Al_2O_3, Fe_2O_3, MgO, and CaO contents increase with decreasing SiO_2 content.

Figure 4.1 shows the elemental composition of volcanic glasses in the form of EDX spectra. The vertical axis (y-axis) shows the relative amount of each atom as a percentage of the total. Although peak height in EDX spectra does not directly yield the actual percentage of each atom in the sample, the EDX spectrum-mimic graphs are useful proxies to compare overall elemental composition quickly.

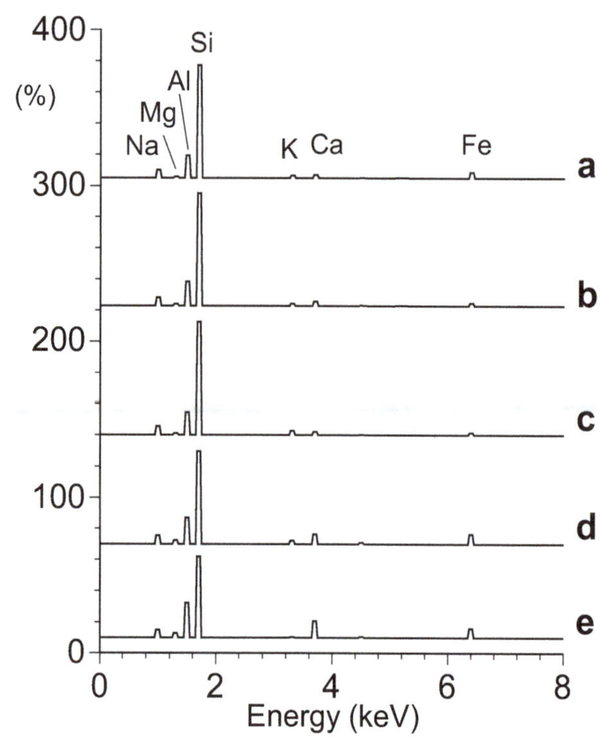

Fig. 4.1 Chemical analyses of volcanic glass collected from volcanic ash of different rock types (Shoji et al. (1975)). (**a**) Towada-a tephra with rhyolite type, (**b**) Chuseri tephra with dacite type, (**c**) Tarumae-a tephra with andesite type, (**d**) Zao-b tephra with basaltic andesite type, (**e**) Hoei tephra with basalt type. The vertical axis (y-axis) shows the percentage (the number of atom) of each element. The horizontal axis (x-axis) shows energy of characteristic X-ray as in an EDX spectrum (EDX spectrum-mimic graphs)

Fig. 4.2 Sponge-like glass particle. (**a**) Optical micrograph, (**b**) SEM image of the dashed area of the sponge-like glass particle, (**c**) EDX spectrum of the dashed area in (**b**). The sponge-like glass particles were collected from the sample shown in Fig. 2.7c (the 2–0.5 mm fraction of Mt. Pinatubo 1991 volcanic ash)

4.2.2 Sponge-Like Volcanic Glass

Figure 4.2 shows an example of sponge-like volcanic glass. Among the three types of noncolored volcanic glasses, the sponge-like volcanic glass is the most common (Shoji 1986). As shown in the scanning electron microscope (SEM) image (Fig. 4.2b), sponge-like volcanic glass is highly vesicular, suggesting strong bubbling in the viscous magma during the eruption. The EDX spectrum in Fig. 4.2c indicates that the Si concentration is high and the Mg and Fe concentrations are low. The sponge-like volcanic glass includes various amounts of phenocrysts. Large sponge-like volcanic glass is synonymous with pumice.

4.2.3 Bubble-Wall Type Volcanic Glass

The bubble-wall type glasses shown in Fig. 4.3 were separated from Kikai-Akahoya (K-Ah) tephra, one of the widely distributed tephra in Japan (Machida 2002b), in Miyazaki Prefecture, Japan. At this site, the K-Ah tephra resides at a depth of several

Fig. 4.3 Bubble-wall type volcanic glass. (**a**) Soil profile having Kikai-Akahoya (K-Ah) tephra, (**b**) close-up of the K-Ah deposit, (**c**) optical micrograph, (**d**) SEM image of bubble-wall type volcanic glass separated from the K-Ah deposit, (**e**) EDX spectrum of a glass particle of dashed square in (**d**)

tens of centimetres, shown with the label "K-Ah" in Fig. 4.3a, b. After H_2O_2 digestion and sieving, bubble-wall type glasses (Fig. 4.3c, d) were handpicked from the sand fraction. The bubble-wall type glasses appear to be formed by pulverization of much large vesicles or bubbles from highly viscous magma. The EDX spectrum (Fig. 4.3e) shows high Si and low Mg and Fe concentrations, and is similar to that shown in Fig. 4.2c.

4.2.4 Fibrous Volcanic Glass

The huge eruption that created the Aira Caldera (northern part of Kagoshima bay, Japan) produced the Ito pyroclastic flow deposit (Aramaki 1984; Baer et al. 1997) (Fig. 4.4a). Fibrous glasses (Fig. 4.4b, c) were handpicked from the pyroclastic flow deposit (Fig. 4.4d) after ultrasonic treatment (Busacca et al. 2001) to remove fine particles. The fibrous glass contains many elongated vesicular pores as shown in Fig. 4.4c. The fibrous glass is a minor type of non-colored volcanic glass. Major constituents of the Ito pyroclastic flow deposit are sponge-like glass, bubble-wall type glass, plagioclase, and quartz, among others (Sonehara 2016). The EDX

Fig. 4.4 Fibrous volcanic glass. (**a**) The Ito pyroclastic flow deposit (29 ka), (**b**) fibrous volcanic glass particles separated from the air-dried Ito pyroclastic flow deposit, (**c**) the SEM image of an edge of fibrous volcanic glass, and (**d**) close-up of the sampled pyroclastic flow deposit of the EDX spectrum of the dashed square

spectrum of the fibrous volcanic glass is close to those shown in Fig. 4.2c and Fig. 4.3e.

4.2.5 *Berry-Like Volcanic Glass*

Mt. Fuji emits scoriaceous tephra (Kobayashi et al. 2007; Miyaji 2002). Soil horizons of A1 to C were formed from Hoei, and horizons of 2A1 from Jogan tephra (Fig. 4.5a). This soil is used as farmland to grow vegetables (Fig. 4.5b). Figure 4.5c shows the 0.5–0.2 mm fraction of the scoriaceous Jogan tephra (2A1 horizon of Fig. 4.5a). This particle-size fraction is colorful, displaying dark-colored scoria (Fig. 4.6), red-colored scoria (Fig. 4.7), olivine (Fig. 2.4), and others.

Figure 4.6 shows an example of a scoria particle or colored volcanic glass. The particle is dark and has many concavities (Fig. 4.6a, b). The EDX spectrum (Fig. 4.6c) of the dashed area (Fig. 4.6b) shows that Mg, Al, Ca, and Fe concentrations are high, and that the Si concentration is low compared with non-colored

Fig. 4.5 Berry-like volcanic glass. (**a**) Scoria deposits from Mt. Fuji, (**b**) landscape of the pedon site, and (**c**) 0.5–0.2 mm fraction of the 2A1 horizon of the scoria deposit

volcanic glasses (Figs. 4.2c, 4.3e, and 4.4c). Many plagioclase crystallites are visible in the magnified SEM image (Fig. 4.6d).

Brownish red particles can be found in Fig. 4.5c, and they are also scoriaceous. Figure 4.7a is a red scoriaceous particle and Fig. 4.7b is the SEM image of Fig. 4.7a that resembles Fig. 4.6b in that it has round concavities. The EDX spectrum (Fig. 4.7c) of the dashed square in Fig. 4.7b is similar to that of the dark-colored scoria particle (Fig. 4.6c) in terms of the high Mg, Al, Ca, and Fe concentration compared with those for Figs. 4.2c, 4.3e, and 4.4c. In the magnified SEM image (Fig. 4.7d), small plagioclases are identified by EDX analyses. The brownish red color of the particle shown in Fig. 4.7a can be explained by high-temperature oxidation of iron.

4.3 Secondary Non-crystalline Inorganic Constituents

Non-crystalline inorganic constituents in this section are newly formed in soil or plants whereas volcanic glasses are of magmatic origin.

Fig. 4.6 A scoria particle separated from Fig. 4.5c. (**a**) Optical photograph, (**b**) SEM image, (**c**) EDX spectrum of the large dashed area shown in (**b**), (**d**) a magnified SEM image of the small dashed area in (**b**)

4.3.1 Allophane and Imogolite

Allophane and imogolite (Yoshinaga 1986; Wada 1989; Harsh et al. 2002) are typical weathering products of volcanic glass under a humid climate and well-drained conditions; they are also found in the spodic horizon soils. Allophane has a spherical structure and is hollow inside with some holes in its wall (Fig. 4.8a). Allophane is a non-crystalline aluminosilicate with a Si:Al atomic ratio of 1:2–1:1. The range of diameters of allophane spherules is 3–5 nm. Imogolite has a tubular structure and a Si:Al atomic ratio of 1:1 (Fig. 4.8b). The diameter of an imogolite tube is 2 nm.

The outside of allophane and imogolite resembles a gibbsite sheet (Fig. 4.8c). The dehydration of three H_2O molecules from the three front OH groups and the three OH groups of $Si(OH)_4$ yields the aluminosilicate wall of allophane (Fig. 4.8c). Because there is no Si-O-Si bond, allophane and imogolite are nesosilicates. The Si-OH groups point towards the inside of the spherules and tubes, as shown in Fig. 4.8a, b, respectively. Allophane having a Si:Al ratio of 1:2 is called Al-rich allophane and is the major type found in Andisols. When the atomic Si:Al ratio of allophane approaches 1, it is considered Si-rich allophane, the Si dimer is thought to

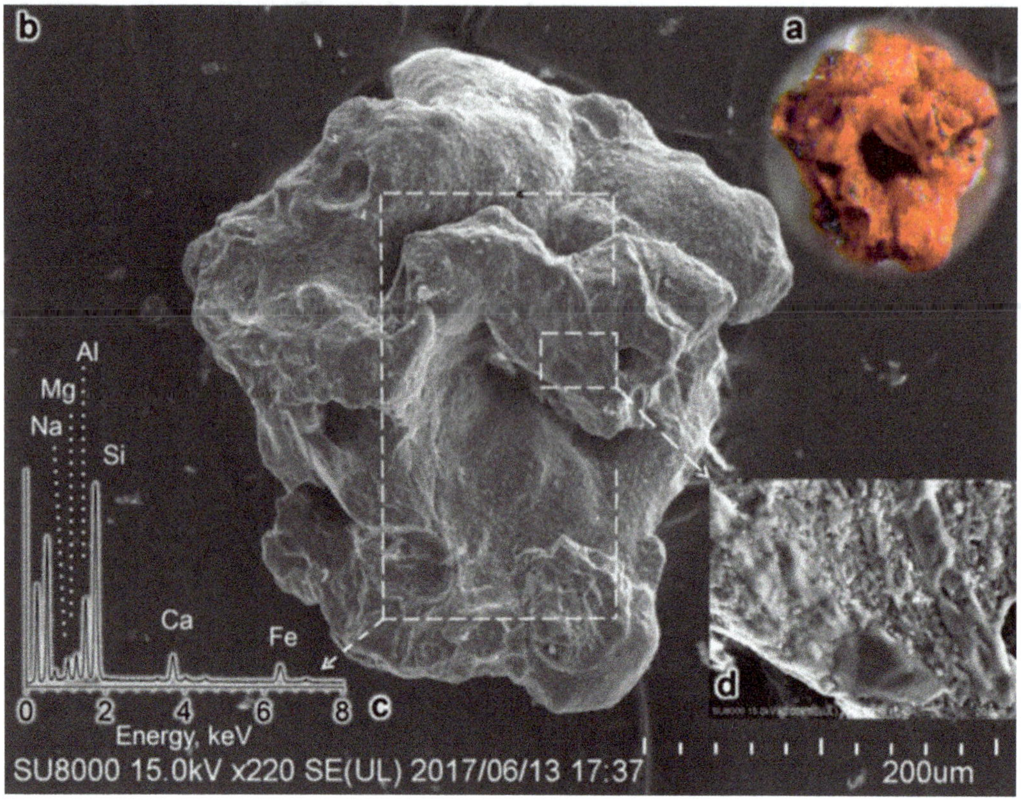

Fig. 4.7 A red scoria particle separated from the sample shown in Fig. 4.5c. (**a**) Optical micrograph (**b**), SEM image, (**c**) EDX spectrum of the large dashed area shown in (**b**), (**d**) a magnified SEM image of the small dashed area in (**b**)

increase. The Si-rich allophane is sometimes found in the weathered pumice layer of the deeper part of Andisol profiles.

Imogolite gel films can be found in field-moist weathered pumice (Fig. 4.9a) taken from the 2Bw1 horizon of the pedon shown in Fig. 4.19. After separation of the film from soil clods and suspending it in pure water followed by ultrasonic treatment, a thin imogolite suspension can be prepared. After spotting the suspension on the micro grid attached copper mesh, a photograph of allophane and imogolite can be obtained using transmission electron microscope (TEM) (Fig. 4.9b). The thin fibrous material visible in Fig. 4.9b is imogolite, and the small distorted circles are allophane. Similar imogolite gel film (Fig. 4.9c) was also found in the 3Bw5 horizon (Fig. 4.19).

As shown in Fig. 4.9a and c, the imogolite gel film exists at the voids among pumice particles, and it contains not only imogolite but also allophane, suggesting that these films are formed through precipitation reactions. Allophanic materials form as altermorphs of vesicle walls inside the pumice particle as shown later in Figs. 4.26 and 4.29. Thus, two types of occurrences are possible for allophane in the weathered pumice.

The allophane and imogolite content can be estimated by oxalate extraction as 7.1 times Si_o, where Si_o denotes the amount of oxalate-extractable Si (Parfitt and Henmi

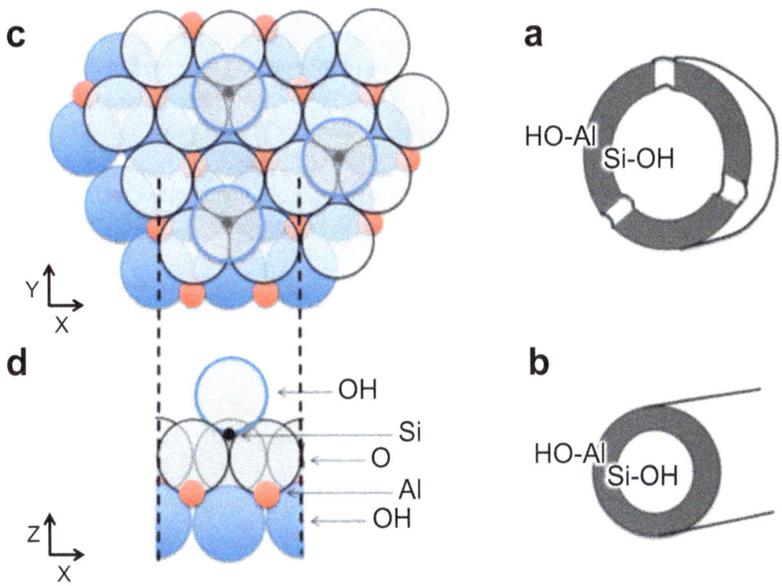

Fig. 4.8 Cross sections of (**a**) allophane and (**b**) imogolite, (**c**) plane view from the inside of a unit particle model, and (**d**) its side view

1982; Parfitt and Wilson 1985). The occurrence of allophane and imogolite is further discussed in Sect. 4.4.

4.3.2 *Laminar Opaline Silica*

Laminar opaline silica or pedogenic opal is a thin, disk-like, or elliptical form of silica (Fig. 4.10b), with diameters ranging from several micrometers to submicrons. Laminar opaline silica is often found in the upper horizons of young volcanic ash within a few thousand years (Shoji and Masui 1969, 1971; Shoji and Saigusa 1978; Henmi and Parfitt 1980). Chemically, laminar opaline silica is non-crystalline silica. Laminar opaline silica is more common in A horizons where Al is complexed with humus and Al-O-Si bonding is inhibited. For example, in the A and Bw horizons of volcanic ash soil, the Si concentration in soil solution is 2–6 mg Si L^{-1} in a humid climate (Ugolini et al. 1988), and Si can become sufficiently concentrated to precipitate silica when evaporation of water from the surface horizon to air is high or, alternately, the surface soil is frozen. Wada and Nagasato (1983) reported formation of silica microplates by freezing a silica solution in the laboratory.

Figure 4.10a shows a profile of Udivitrand in Hokkaido, Japan. In the clay fraction of the uppermost A horizon, laminar opaline silica particles can be easily found. The EDX spectrum of laminar opaline silica is highly dominated by Si (Fig. 4.10c).

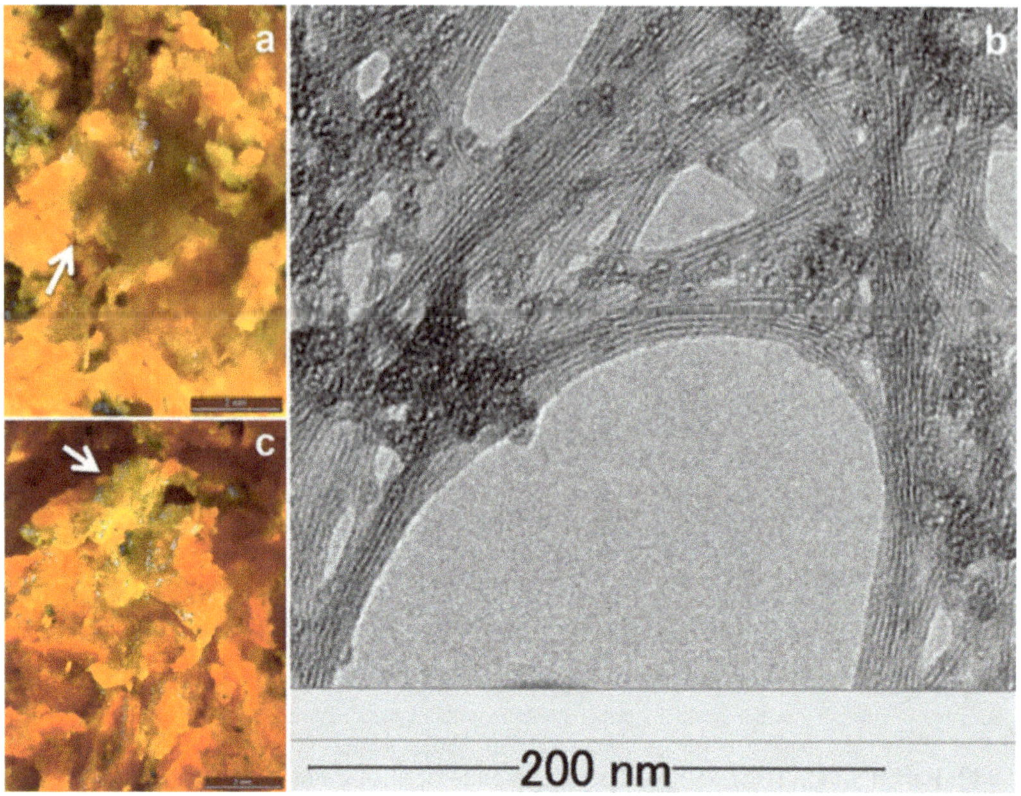

Fig. 4.9 Allophane and imogolite. **(a)** A gel film (white arrow) in the 2Bw1 horizon (Fig. 4.19a), **(b)** TEM image of allophane and imogolite in the gel film, **(c)** similar gel film (white arrow) found in the 3Bw5 horizon (Fig. 4.19a)

4.3.3 Phytoliths

Phytoliths, or plant opals, are a type of biogenic opals (Wilding et al. 1977; Drees et al. 1989). Phytoliths are typically found in the particle size fraction of 5–50 μm and are larger than laminar opaline silica (Fig. 4.10b). Phytoliths are abundant in A or buried A horizons. Phytoliths are non-crystalline silica, formed in plant cells. As the morphological form of a phytolith is specific to plant species to some extent, the form of phytoliths in a buried A horizon can be used to estimate the paleovegetation of the buried A horizon (Kondo et al. 1988). Phytoliths possibly benefit plants by contributing to Si nutrient cycling.

Figure 4.11a shows an example of the Udand profile, which has abundant phytoliths in the A1 horizon. Figure 4.11c is an optical micrograph of the 0.05–0.02 mm fraction, and particles indicated by white arrows can be judged as phyoliths according to their morphological characteristics.

Biogenic opals, laminar opaline silica, and volcanic glasses are non-crystalline. Morphological characteristics are used to identify non-crystalline inorganic constituents. Although XRD is a powerful tool for the identification of crystalline constituents, it is not effective for non-crystalline inorganic constituents. Elemental composition is an useful tool for identification of non-crystalline inorganic

Fig. 4.10 Occurrence of laminar opaline silica. (**a**) Soil profile of Udivitrand which includes opaline silica at the A horizon, (**b**) SEM image of the clay fraction containing thin disc-like opaline silica, (**c**) EDX spectrum of laminar opaline silica

constituents. EDX is effective for the determination of the approximate elemental composition of particles in soil.

In order to examine the andic soil properties in USDA Soil Taxonomy and andic properties in World Reference Base for Soil Resources, glass counting is needed for the young soils (Eden 1992). In glass counting, the 2–0.02mm fraction is used and a distinction between volcanic glass and phytoliths is necessary. Subdivision of the 2–0.02mm fraction into 2–0.2, 0.2–0.05 and 0.05–0.02mm fractions is effective in allowing each particle to be identified. Fractionation of these particle-size fractions should be done after H_2O_2 digestion, dithionite-citrate-bicarbonate treatment, and wet sieving. If data on the smaller particle-size fraction are needed, the 0.02–0.005-mm fraction can be prepared, removing the fraction less than 0.005-mm fraction by a dispersion–sedimentation and siphoning procedure. For quantitative purpose, eight repetitions of this procedure must be performed.

A few characteristics of the 2–0.005mm particle-size fraction can be noted. Phytoliths are not included in the 2–0.2mm fraction, and are not prevalent in the B and C horizon soils. Volcanic glasses are present in all size fractions, although the abundance of volcanic glass in each particle-size fraction changes depending on the individual sample (Yamada and Shoji 1975).

Figure 4.11c shows an optical micrograph of the 0.05–0.02mm fraction sampled from A1 horizon of Fig. 4.11a. From the morphological characteristics of the

Fig. 4.11 Occurrence of phytoliths. (**a**) Udand profile displaying abundant phytoliths in the 0.05–0.02 mm fraction of the A1 horizon, (**b**) present-day vegetation, (**c**) optical micrograph of the 0.05–0.02 mm fraction

particles, the three particles indicated by white arrows were identified as phytoliths. However, identification may be rendered difficult owing to weathering or the presence of composite particles containing different materials. In that case, SEM image (Fig. 4.12a), element maps (Fig. 4.12b, c), and micrographs (Fig. 4.12d) are effective tools for distinguishing among volcanic glass, phytoliths, and crystalline minerals. An important point is that the same sample must be used for both polarizing microscope and SEM-EDX observations. For a detailed procedure, see footnote[1].

[1]The first step is to prepare a sample for both polarizing microscope and SEM observations. To do so, spread sample particles on a glass slide without overlapping. For the 0.05–0.02 and 0.02–0.005 mm fractions, place about 2–3 mg of air-dried powder on a glass slide by hand, and perform the following procedure: tilt the slide, holding one side of the slide, tap the other side of the slide on a table so that the particles gradually slide down and spread evenly on the slide.

Use one side of transparent double-sided sticky tape to pick up the particles from the glass slide, and stick the other side of the tape to a cover glass. A disk-shaped cover glass is more convenient for handling than square one. Put the cover glass on a glass slide, and then put the slide on the rotating stage of a polarizing microscope (Lynn et al. 2008). Take photographs with plain and crossed polarizers at least 2 different angles separated by 45° to distinguish isotropic (non-crystalline) and anisotropic (crystalline) particles (Fig. 4.12d). Use the photographed particles for the next step.

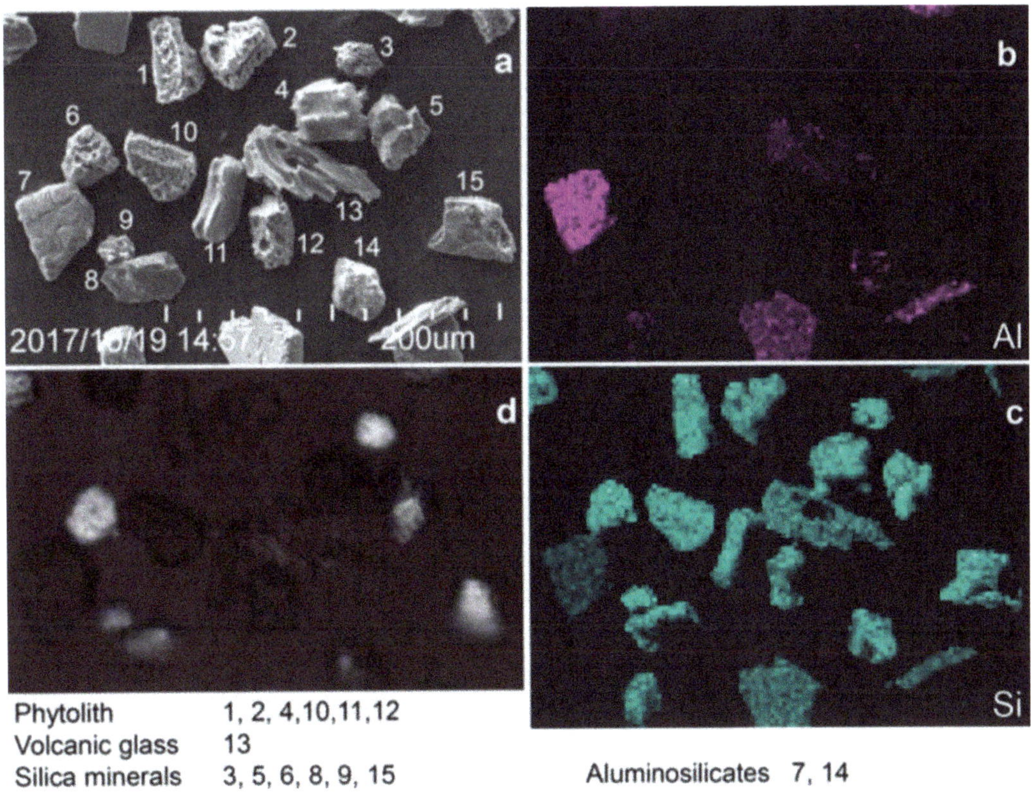

Fig. 4.12 Distinction of inorganic particles. (**a**) SEM image, (**b**) Al and (**c**) Si element maps, (**d**) photograph using crossed polarizers of the lower right area of Fig. 4.11c

Only the wholly visible particles in Fig. 4.12 (numbered particles 1 through 15) were targeted for interpretation. Figure 4.12b, c are element maps of Al and Si, respectively. Particles 7, 13, and 14 are seen to contain Al. All the particles contain Si. Particles 1 to 6, 8 to 12, and 15 contain few elements other than Si, suggesting that they are silicas. Among these silica particles, a micrograph with crossed polarizers (Lynn et al. 2008) (Fig. 4.12d) suggests that particles 3, 5, 6, 8, 9, 14, and 15 are anisotropic, which indicates that they are crystalline silica minerals. The same results were obtained after turning the rotating stage of the polarizing microscope. Other silica particles (1, 2, 4, 10, 11, and 12) are isotropic, suggesting that they are phytoliths. Particle 13 is identified as volcanic glass from its vesicular morphological properties in addition to its elemental composition (Fig. 4.12b, c) containing both Si and Al. Particles 7 and 14 appear to be aluminosilicates. Although

---

The second step is to observe the morphological properties of each particle using high-resolution SEM imaging (Fig. 4.12a). Coating with vacuum-evaporated carbon is desirable for obtaining EDX data. The third step is to obtain element maps (Na, Mg, Al, Si, P, K, Ca, Mn, Ti, and Fe) of the same SEM image. Free software packages are available for handling rotating polar microscope images, for overlaying element maps with other maps and photographs, and for counting particles. Among the isotropic particles, the particles having only Si and no other elements are plant opals. Among the isotropic particles, the particles having Al are volcanic glass. Examine the EDX spectrum of the particle in order to minimize identification errors.

particle 7 is isotropic, its Al content is too high to allow for identification as a volcanic glass.

Diatoms, which are also biogenic opals, are also found in the clay or silt fractions of A horizon soils. Diatoms can be identified by their characteristic frustule. As laminar opaline silica is almost always smaller than 5 μm, it does not affect the glass counting of the 2–0.02 mm fraction.

Thus, by using both polarizing microscope and SEM-EDX, the ability to identify non-crystalline soil constituents such as volcanic glass and phytoliths is greatly enhanced. Remaining issues include a few phytoliths and diatoms that are weakly anisotropic, and crystalline particles completely covered by volcanic glass. Although some iron minerals are not transparent under optical microscope, they can be identified using an Fe element map.

4.3.4 *Al-Humus Complex*

Al-humus complex (Al-humus) is one of the active Al materials in Andisols and Spodosols. Al-humus was recently reviewed by Takahashi and Dahlgren (2016). This section is focused on Al-humus in Andisols, because in these soils Al-humus is much less mobile, whereas that in Spodosols appears more mobile and is related to eluviation–illuviation processes (Ito et al. 1991). The justification for covering Al-humus despite its partially organic structure is that (i) it is highly reactive with phosphate, (ii) Al-humus formation appears to be competitive with allophane and imogolite in Andisols depending on the pH and organic matter content, and (iii) it is the major active Al form in the A horizons of non-allophanic Andisols.

The amount of Al in Al-humus is typically measured as the Al extractable by pyrophosphate extraction (Al_p) (Dahlgren 1994). Al-humus cannot be chemically isolated from Andisols, but its structure is roughly estimated from cation exchange capacity (CEC), Al_p, functional group analysis of humic acid (Fig. 4.13), and OH^- release through the reaction with F^-. The presence of hydroxyl groups in Al-humus

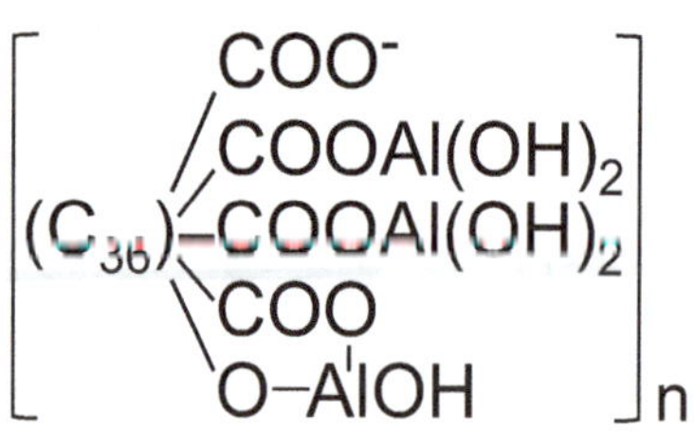

Fig. 4.13 The approximate formula of an Al-humus complex at pH 7. Of every 40 organic carbon atoms, four belong to carboxyl groups in A-type humic acid (Yonebayashi and Hattori 1988), three of which are complexed with Al (Shoji et al. 1993), and one of which will dissociate to have a negative charge. The amount of negative charge was derived from the slope of the regression line between organic C content and CEC

is suggested by the high pH (NaF) of non-allophanic Andisols (Shoji et al. 1985). Figure 4.13 was constructed using data from A horizon soils of typical non-allophanic Andisols excluding the uppermost A horizons with fresh organic matter, which cannot yet be Al-humus.

4.4 Andisols: Soils Dominated by Non-crystalline Inorganic Constituents

Non-crystalline inorganic constituents are important to both the parent materials and soil formation products of Andisols. The major parent material of Andisols is volcanic ash, and the major constituent of the volcanic ash is volcanic glass (Fig. 4.14 and Table 4.2). Major soil formation products in Andisols are allophane, imogolite, and Al-humus complexes, especially under humid climate and good drainage conditions. Kaolin minerals are also present in the soil formation products in the lower horizons of Udands, in Udands under poor drainage, and in Andisol under semi-dry climate. In this section, the processes of Andisol formation are introduced (Shoji et al. 1993; Arnalds et al. 2007; McDaniel et al. 2012).

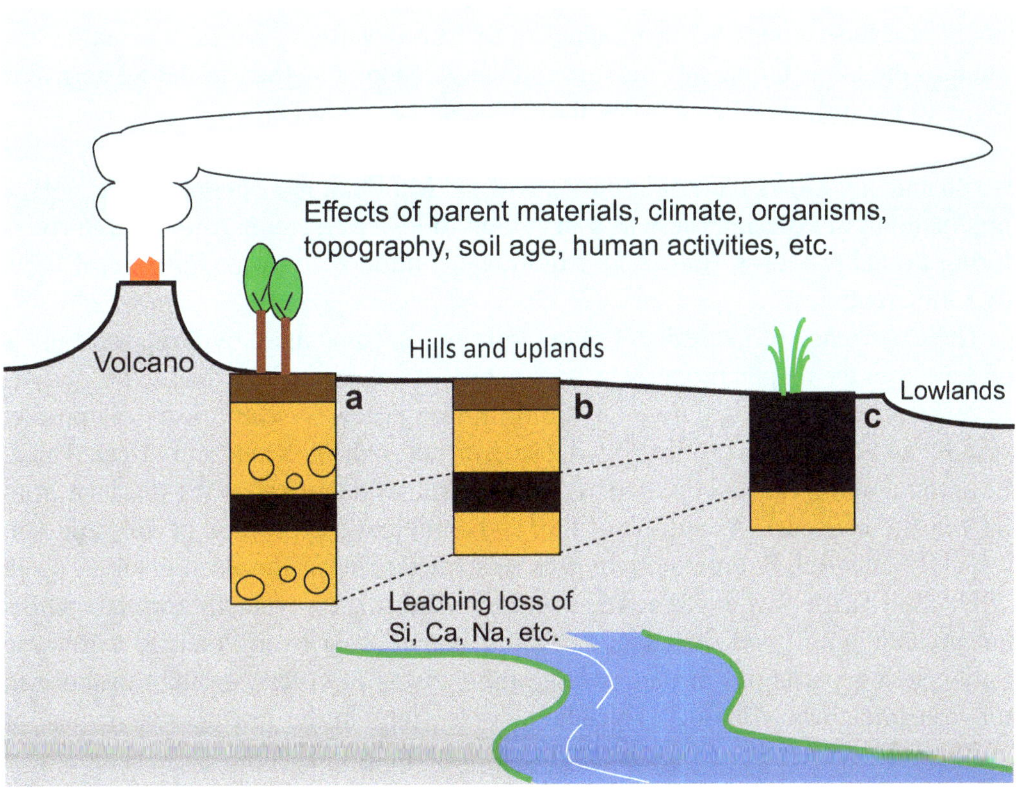

Fig. 4.14 Schematic of Andisol formation from volcanic ash. Different profiles (**a**, **b**, and **c**) are formed depending on the soil forming factors

Table 4.2 The Si/Al atomic ratio of major constituents in volcanic ash soils

	Constituents in volcanic ash soils	Si/Al atomic ratio
Parent materials	Non-colored volcanic glass	≈5.0[a]
	Colored volcanic glass	≈2.4[a]
Soil formation products	Halloysite	1
	Allophane	≈0.5-1
	Imogolite	0.5
	Al-humus	≈0

[a]The Si/Al atomic ratios of volcanic glasses were calculated from Shoji (1986). Others are theoretical values

The distribution of Andisols is largely determined by the locations of volcanic regions. The major volcanic regions in the world are the Circum-Pacific volcanic zone, the Hawaiian Islands, the Sumatra-Java-Lesser Sunda Islands, Eastern Africa, the Mediterranean area, Iceland, the Azores, and the Canary Islands (Sievert et al. 2010; Machida 2002a). In the case of huge volcanic eruptions, which occur intermittently, interspersed with dormant periods of hundreds or thousands of years, a large amount of volcanic ash affects wide areas (Machida 2002a). Volcanic ash deposits in these areas are important parent materials of soils.

Figure 4.14 shows a schematic of Andisol formation from volcanic ash. Highly explosive eruptions are caused by viscous, Si-rich magma, which form non-colored volcanic glass. In the mid-latitudes, volcanic ash is mainly distributed on the eastern side of volcanoes owing to the influence of the strong westerly winds, whereas at low latitudes the ash is found either all around the volcano or slightly to the western side owing to the influence of the weak trade winds.

Volcanic ash deposits on the slopes of volcanoes are further transported by water as volcanic mudflows (lahar). In the case of the Mt. Pinatubo ash deposit in 1991, a large amount of volcanic ash was transported to lower elevations along large rivers during the rainy season (Fig. 2.7). Mudflows continued to occur for several years after the eruption.

The occurrence of intermittent huge eruptions accompanied by large ash falls is evident from the thaptic property in the vicinity of a volcano. The thaptic property is the existence of a buried Andisol profile under a new volcanic ash soil profile, forming a multisequum profile (Fig. 4.14a, b). Within the volcanic ash-affected area, the amount of ash fall and the size of the ash particles decreases as the distance from the volcano increases. As a result of thin and cumulative deposition of volcanic ash, a thick humus-rich A horizon, which is a characteristic of the Pachic subgroup in USDA Soil Taxonomy, is formed (Fig. 4.14c). Andisols are typically formed on hills and uplands with good drainage. Leaching losses of Si from volcanic ash cause higher Si concentrations in the rivers in the vicinity of volcanoes than in those in non-volcanic areas. Although volcanic ash also falls in the lowlands, marshes, or basins with restricted drainage areas, the weathering of volcanic glass is slow and halloysite is formed, possibly because high Si concentrations are maintained in the soil water. The formation of smectite from volcanic glass is also reported under hydrothermal conditions in the laboratory (Tomita et al. 1993; Cuadros et al. 1999).

Although humus-rich horizons are generally dark-colored, highly black colored A horizons are often found beneath grass vegetation. The grass vegetation was maintained even under humid climates through burning by ancient people. Under forest vegetation, the color of A horizons is dark brown. The difference in color between these two types of A horizons is due to the differences in their humic acid types. The humic acid, under grass vegetation is A-type, which is highly humified and rich in aromatic groups whereas that under forest vegetation is either P-type or B-type. The A-type humic acid is separated from other types of humic acid by having a melanic index lower than 1.7, whereas the other types of humic acid have melanic indices of 1.7 or higher (Shoji et al. 1993).

Andisols are characterized by high active Al and Fe contents. The active Al materials are allophane, imogolite, and Al-humus, and the active Fe material is ferrihydrite. These materials are mainly formed from volcanic glass. The typical Si:Al atomic ratios of non-colored and colored volcanic glasses are 5.0 and 2.4, respectively. These values are significantly higher than those for soil formation products such as kaolin minerals (often halloysite), allophane, imogolite, and Al-humus, as shown in Table 4.2. Morphological changes of volcanic glasses accompanying these changes in elemental composition were examined using polished sections of new pumice particles, partially weathered pumice particles and soil clods from Udands.

4.4.1 *Fresh Pumice Particle*

A sample of fresh pumice was obtained from the 1991 Mt. Pinatubo tephra. Figure 4.15a shows an optical micrograph of the polished section. The white part is sponge-like volcanic glass with inclusion of feldspar, quartz, and other particles. Figure 4.15b shows an EDX spectrum of a glassy area of the particle shown by a dashed square (b) in Fig. 4.15a, and it is close to the typical non-colored volcanic glass (Fig. 4.2c). Magnifying the rim of the pumice, fine particles of 10 μm or less are found in the open cavities, and these are also non-colored volcanic glasses as shown by the similar EDX spectrum (Fig. 4.15e) to that for Fig. 4.15b. Figure 4.15e may correspond to the start of the micromass coating formation (Stoops 2007). Highly vesicular characteristics similar to Fig. 4.2b are seen inside the pumice.

Figure 4.15f, g are the element maps of Al and Si, respectively. The cyan color used to indicate Si displays its second highest intensity for the major and glassy part of the pumice particle. The highest color intensity of Si may indicate that some quartz was included in the pumice. The magenta color chosen for Al is strongest for feldspar particles, and the intensity of the magenta color for the volcanic glass is lower than the cyan color intensity as suggested from the EDX spectra shown in Fig. 4.15b, e.

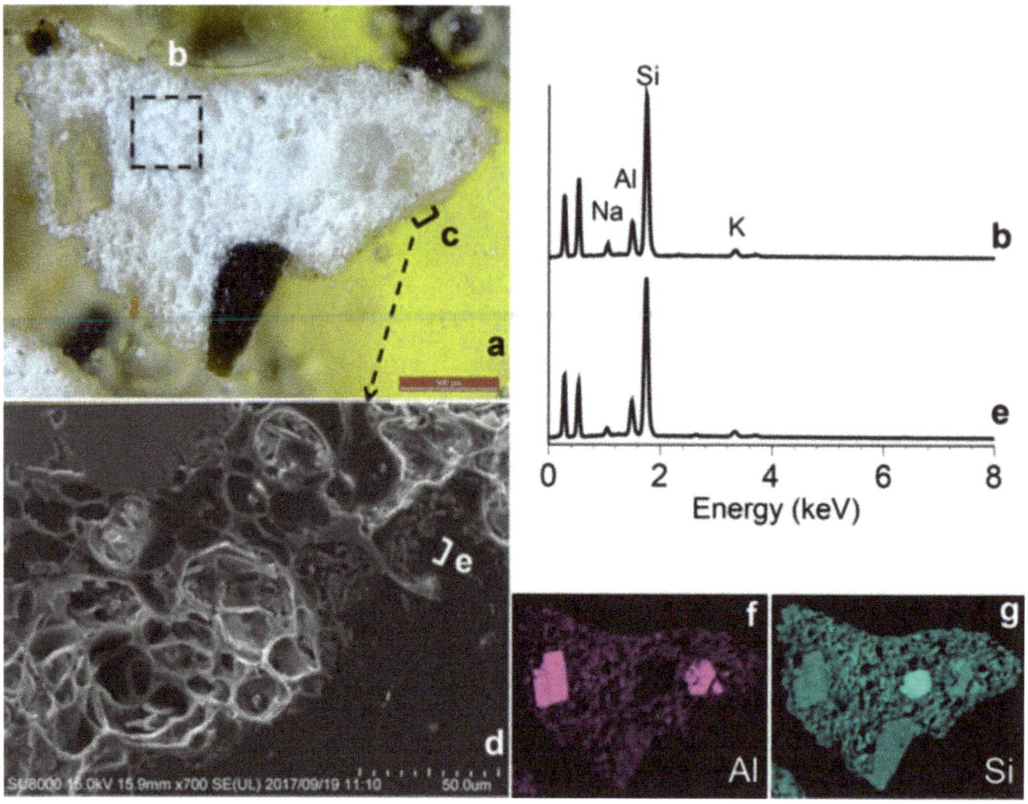

Fig. 4.15 Fresh pumice. (**a**) Polished section of a pumice particle obtained from the 1991 Mt. Pinatubo volcanic ash deposit, (**b**) EDX spectra of the selected areas (dashed square (**b**) in (**a**)), (**c**) very small particles, (**d**) SEM image of the selected area (**c**) of the pumice particle, (**e**) EDX spectrum of a fine particle from the area (e), (**f** and **g**) Al and Si element maps of the pumice particle, respectively

4.4.2 Partially Weathered Pumice Particle

In the Central Plain of Luzon, the ash from the 1991 eruption (C horizon) deposited on the previous volcanic ash soil (2A1 or below) (Fig. 4.16a). A partially weathered pumice from the 2Bw horizon was sampled and a polished section (Fig. 4.16b) was prepared. The particle is brownish all the way through, suggesting oxidation and weathering of iron within the pumice. The primary land use type around the pedon is pasture (Fig. 4.16c), and the grass vegetation on the new ash had recovered by 1993, because the depth of the new ash at this site was only 13 cm.

Figure 4.17a, b shows element maps for Fig. 4.16b. The Al concentration, indicated by the magenta color, appears to be high for the phenocrysts, suggesting that these may be feldspars. Phenocrysts aside, the Al concentration at the rim of the pumice particle appears to be higher than the sponge-like inside of the pumice particle. The intensity of the cyan color, which indicates Si, appears nearly identical to the fresh pumice particle. Figure 4.17c shows a magnified SEM image of the upper part of the pumice. Sponge-like or highly vesicular structures can be found

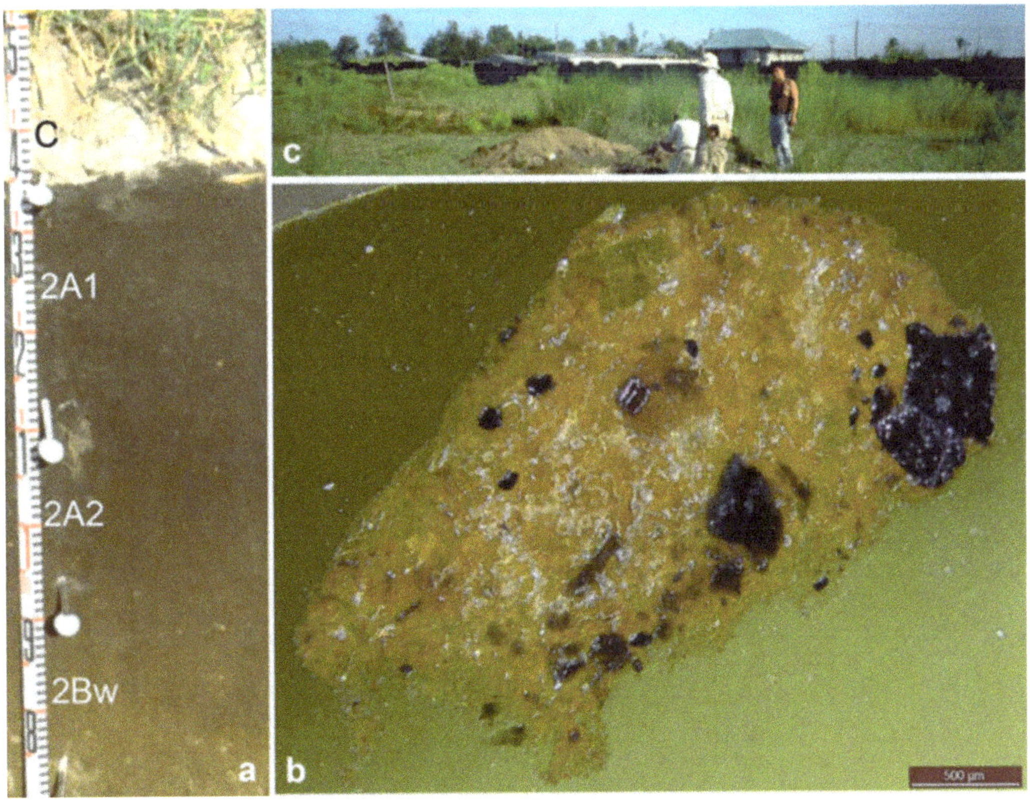

Fig. 4.16 Partially weathered pumice. (**a**) Soil profile, (**b**) polished section of a weathered pumice obtained from 2Bw horizon of (**a**), (**c**) a landscape around the soil profile

inside the pumice. At the rim of the pumice, small particles, similar to those seen in Fig. 4.15e, are found. Several large pores display vacant parts where the penetration of resin was insufficient. The area shown by the dashed square (Fig. 4.17d) was further magnified in Fig. 4.18.

Figure 4.18a, b shows element maps for Si and Al, respectively. The intensity of the cyan color, which indicates Si concentration, is somewhat weak at the rim, although similar low intensities are found in parts of the interior. The intensity of the magenta color, which indicates Al concentration, is especially high at the rim, for example at the site indicated by the white arrow labeled (d) in the SEM image (Fig. 4.18c), where Si concentration appears low. In contrast, the Si concentration still appears to be quite high at the site indicated by the white arrow labeled (e) in Fig. 4.18c. These observations can be confirmed by the EDX spectra (lower right of Fig. 4.18) of the small parts shown by white arrows labeled (d) and (e) in Fig. 4.18c. The EDX spectrum (d) shows high Al and low Si concentrations, suggesting the existence of allophane and imogolite, whereas the EDX spectrum (e) is close to that of the non-colored volcanic glass (Fig. 4.15e). Thus, consistent with the brownish color of the pumice in Fig. 4.16b, partial weathering of the pumice particle is shown by the SEM-EDX analyses. In this example, weathering appears most intensive at the rim sites of the pumice particle.

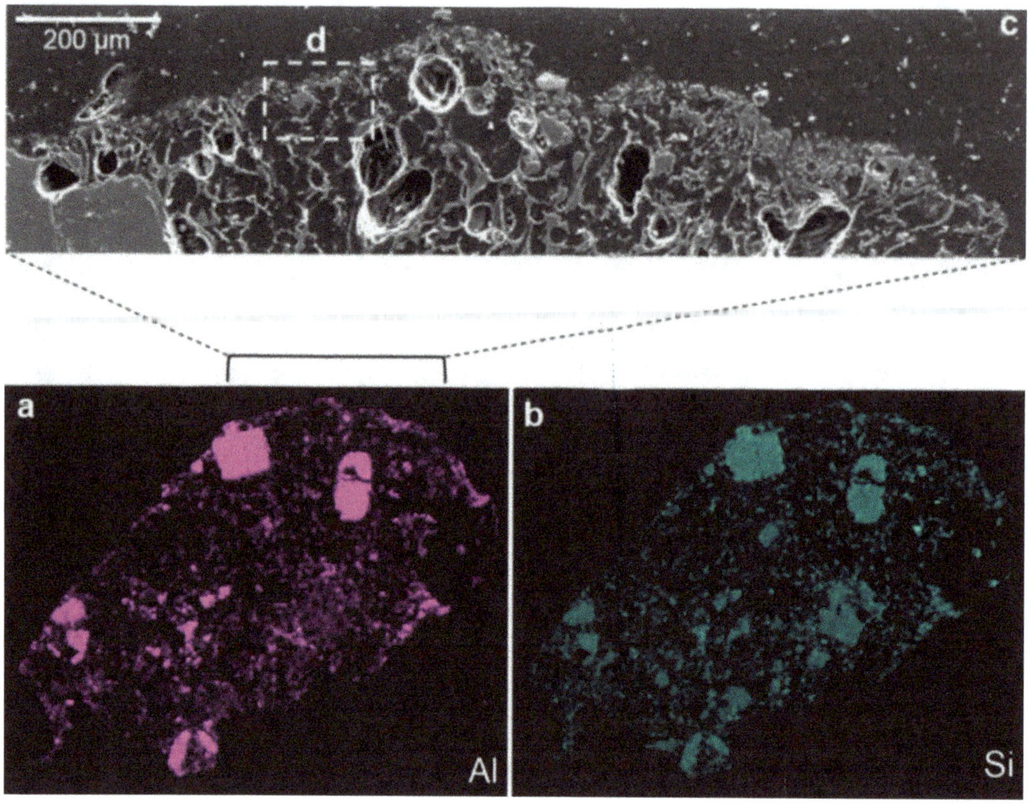

Fig. 4.17 The SEM-EDX analyses of partially weathered pumice. (**a** and **b**) Al and Si element maps, respectively, of the polished section in Fig. 4.16b, (**c**) magnified SEM image of the upper selected part of the pumice particle

4.4.3 A Horizon Soil with Andic Soil Properties

The next example is a matured Andisol including a humus-rich A horizon and highly weathered Bw horizons. Figure 4.19a shows a Pachic Melanudand profile. The profile has thick black A horizons, yellow 2Bw horizons (Nantai-Shichihonzakura (Nt-S) tephra), and reddish brown 3Bw horizons (Nantai-Imaichi (Nt-I) tephra, Ishizaki et al. 2017). Both Nt-I and Nt-S were ejected from Mt. Nantai, Japan. As no other layer is found between the two, the interval between Nt-I and Nt-S deposition is estimated to be short. The age of Na-I is estimated to be ca. 17 ka (Ishizaki et al. 2017 and references therein). By definition, the bulk density of matured Andisols is 0.90 Mg m^{-3} or less. Bulk density values of matured Andisols are as low as 0.5 or 0.6 Mg m^{-3}.

The micromorphological and chemical properties of the A3, 2Bw1, and 3Bw5 horizons were examined using thin sections and polished sections. The vegetation of the sampled site is Quercus Serrata and undergrowth (Otowa and Shoji 1987) (Fig. 4.19b). Thin sections of the A3 horizon (Fig. 4.19c) showed porous and granular microstructure (Stoops 2007) similar to those reported by Kawai (1969). In the 2Bw1 horizon (Fig. 4.19d), a coating of fine, light brown material is probably a more weathered part of pumice particles. Vesicular structure remains inside the

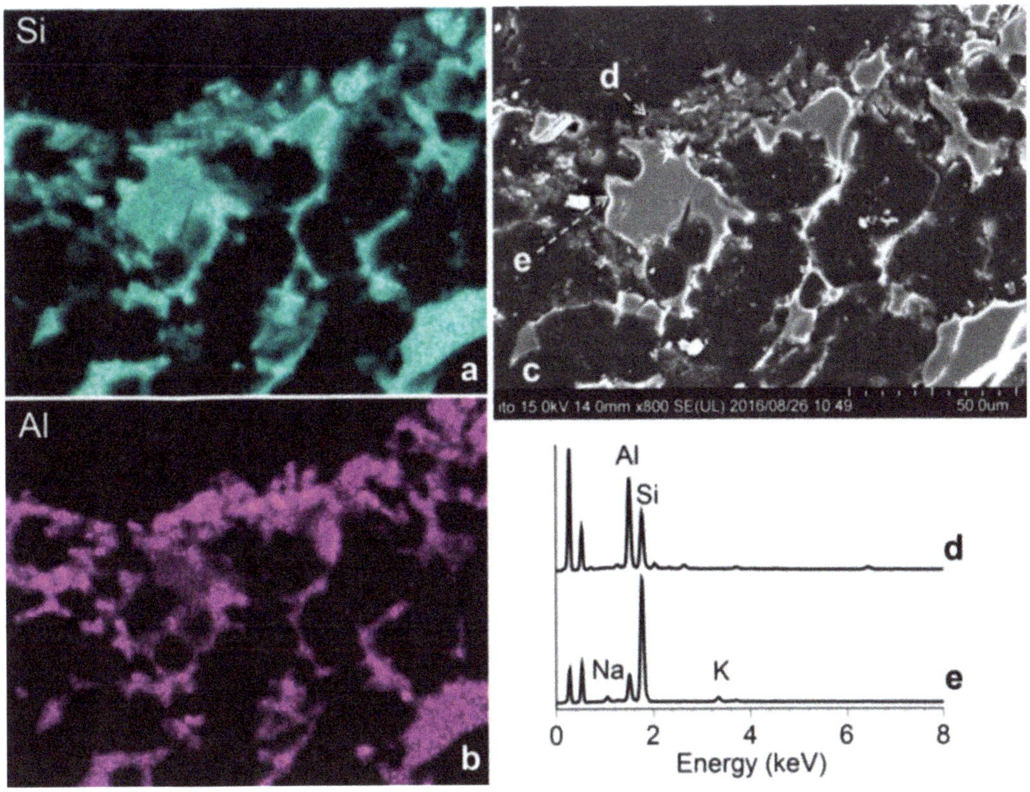

Fig. 4.18 Magnified SEM-EDX analyses of partially weathered pumice. (**a** and **b**) Si and Al element maps, respectively, (**c**) SEM image of the selected area (**d**) of Fig. 4.17c, (**d** and **e**) EDX spectra (lower right) of the spots (**d**) and (**e**) indicated by white arrows in (**c**)

pumice. The light gray parts of Fig. 4.19d are voids. In the 3Bw5 horizon (4.19e), small light brown and orange parts are found within the weathered pumice. These parts are similar in appearance to the orange-reddish altermorph reported for the Bw3 horizon of the Andosol of Tenerife by Stoops (2007). The light gray parts are also voids in Fig. 4.19e. The micromass of these three thin sections, Fig. 4.19c–e, showed undifferentiated b-fabric between crossed polarizers, indicating that isotropic or non-crystalline materials are dominant in the fine materials of these horizons.

SEM-EDX analyses of polished sections further reveal the properties of the A3, 2Bw1, and 3Bw5 horizons. Figure 4.20a shows an optical microscope photograph of the polished section prepared from a clod of the A3 horizon (Fig. 4.19a). The dashed square (Fig. 4.20b) in the upper left corner of Fig. 4.20a was magnified in Fig. 4.20c. Weathered minerals and other particles appear in the dark-colored fine materials with the humus. Figure 4.20d, e, the same area as in Fig. 4.20c, show element maps for Al (magenta) and Si (cyan), respectively. Comparing the color intensity, Al concentration is higher than Si for all of the fine materials except for the coarse particles. This result is highly contrasting with Fig. 4.15f, g where the Si concentration is higher than the Al concentration. The element maps of Fig. 4.20d and e are the result of Andisol formation, consisting of changes in the Si-rich parent material of volcanic glass to form Al-rich allophane, imogolite and Al-humus (Table 4.2). The concentrations of oxalate-extractable Al (Al_o), Al_p, and Si_o in horizon are 7.6, 2.2, and

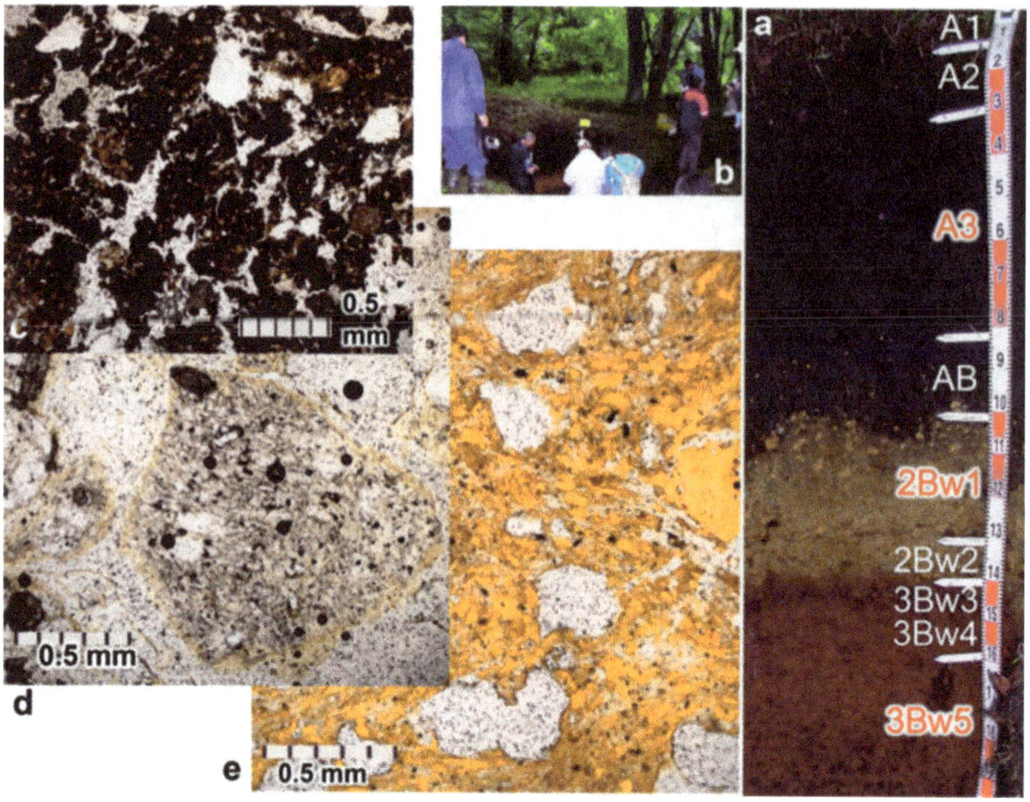

Fig. 4.19 Thin sections of a matured Andisol. (**a**) Pachic melanudand profile from Kiwadashima, Tochigi, Japan, (**b**) landscape around the soil profile, (**c**, **d**, and **e**) images of the thin section of soil clod from the A3 horizon, 2Bw1 horizon, 3Bw5 horizon in plain-polarized light, respectively

2.5%, respectively, indicating that the A3 horizon is highly weathered and allophanic, and that it contains Al-humus complexes. Nevertheless, there are still coarse silica particles that are probably cristobalite, quartz, or phytoliths. In Fig. 4.20c, a few reticulated patterns were observed, and one of them (highlighted by dashed square (f)) was magnified in Fig. 4.21a.

Figure 4.21a shows a reticulated pattern of Fig. 4.20f with magnification, and the EDX spectrum of the dashed area is highly dominated by Al (Fig. 4.21b), whereas the EDX spectrum of the fine materials near the reticulated pattern suggests allophanic material (Fig. 4.21c). The Al element map of the reticulated area (Fig. 4.21d) shows that the Al concentration is nearly homogeneous with little Si present (Fig. 4.21e). These results suggest that the reticulated material is a sclerotia grain.

Sclerotia grains separated from the A3 horizon soil were examined. Water was added to the air-dried soil and ultrasonic treatment was carried out for 10 min. After allowing the suspension to stand for several minutes, dark spherical particles floating on the surface of water were ladled out on a filter paper and air-dried. Figure 4.22a shows the spherical particles separated from the A3 horizon soil, and Fig. 4.22b is the magnified SEM image of the particle. Further magnification of the dashed square in Fig. 4.22b gives Fig. 4.22c, where small holes appear. The hole is one of the characteristics of the sclerotia grain (Watanabe et al. 2001, 2002, 2004, 2007). The

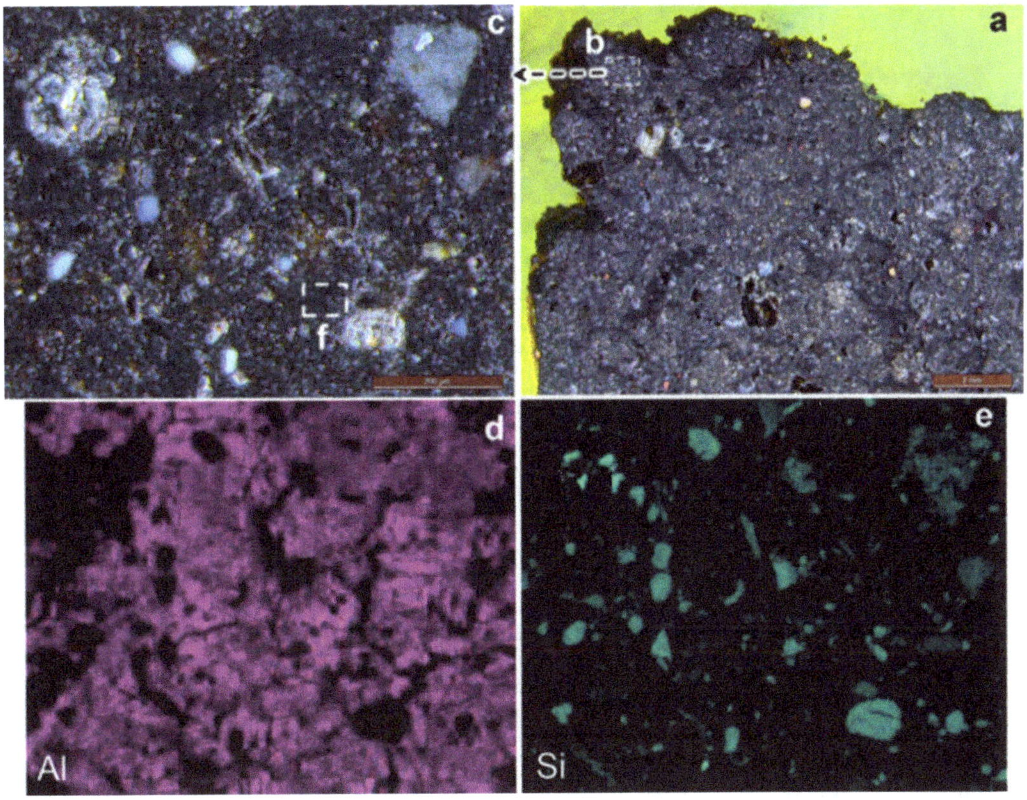

Fig. 4.20 Polished section of A horizon soil. (**a**) Polished section of a clod obtained from A3 horizon of Fig. 4.19a, (**c**) magnification of the dashed square (**b**) of (**a**), (**d** and **e**) Al and Si element maps of (**c**), respectively

EDX spectrum of the dashed square in Fig. 4.22d shows that the outside of the sclerotia grain is Al-rich (Fig. 4.22e)

The largest particle shown in Fig. 4.22a was broken to examine the inside of the particle, of which the SEM image is shown in Fig. 4.23a. The inside of the particle displays a reticulated structure. A magnified SEM image (Fig. 4.23b) of the dashed square in Fig. 4.23a shows that each concavity has a few small holes, which is characteristic of sclerotia grains (Watanabe et al. 2002). The EDX spectrum obtained from the whole of Fig. 4.23b shows that the inside of the particle is also Al-rich. Hence, the Al-rich reticulated structure observed in Fig. 4.21 is the sclerotia grain. Sclerotia grains contains green pigment compounds related to perylene quinone (Kumada and Hurst 1967). These green pigment compounds are typically found in soils containing P-type humic acid. Except the sclerotia grains, a special distribution pattern for Al is not found in Fig. 4.20a.

The formation of Al-humus, phytoliths, sclerotia grains, and diatoms results from biological activities in the A horizons. The formation of laminar opaline silica may be facilitated by the formation of Al-humus. Allophane, imogolite, and ferrihydrite are formed from inorganic parent materials not only in the A horizons but also in the Bw horizons of Andisols.

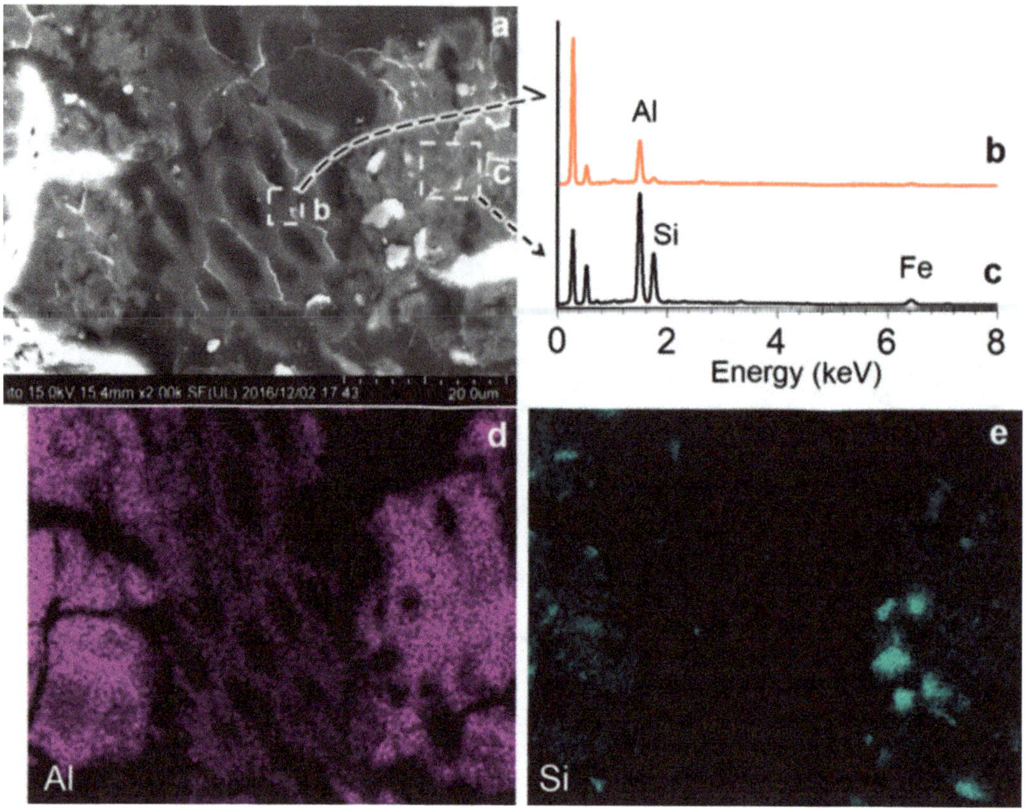

Fig. 4.21 SEM-EDX analyses of a reticulated pattern found in the polished section (Fig. 4.20f). (**a**) SEM image of Fig. 4.20f, (**b** and **c**) EDX spectra of the selected areas (**b**) and (**c**) in (**a**), respectively. (**d** and **e**) Al and Si element maps of (**a**), respectively

4.4.4 B Horizon Soil with Andic Soil Properties

The 2Bw1 horizon consists of weathered Nt-S tephra. Figure 4.24a shows part of a polished section prepared from a clod consisting of weathered pumice. Although the color is yellowish brown, the inside of the pumice retains its highly vesicular structure. The outside of pumice is browner than the inside. The Al element map (Fig. 4.24d) shows that the Al concentration is high along the rim of weathered pumice. The EDX spectrum (Fig. 4.24e) suggests that the dashed square e in the Fig. 4.24d is an allophanic material. Regarding the inside of the weathered pumice, the Al concentration also appears to be increasing. Al and Si element maps were also examined for the dashed square (c) in Fig. 4.24a.

Figure 4.25 shows a magnified SEM image of the dashed square Fig. 4.24c. Weathered phenocrysts and many small bubbles, probably owing to inadequate penetration of resin to the small vesicular pores, are observed. Nevertheless, comparing the color intensity of the Al and Si element maps, Fig. 4.25b, c, respectively, the concentration of Al appears to be higher in the vesicular or sponge-like part than that of Si. To further examine the distribution of Al and Si, the dashed squares (d) and (e) in Fig. 4.25a were magnified and are shown in Figs. 4.26 and 4.27, respectively.

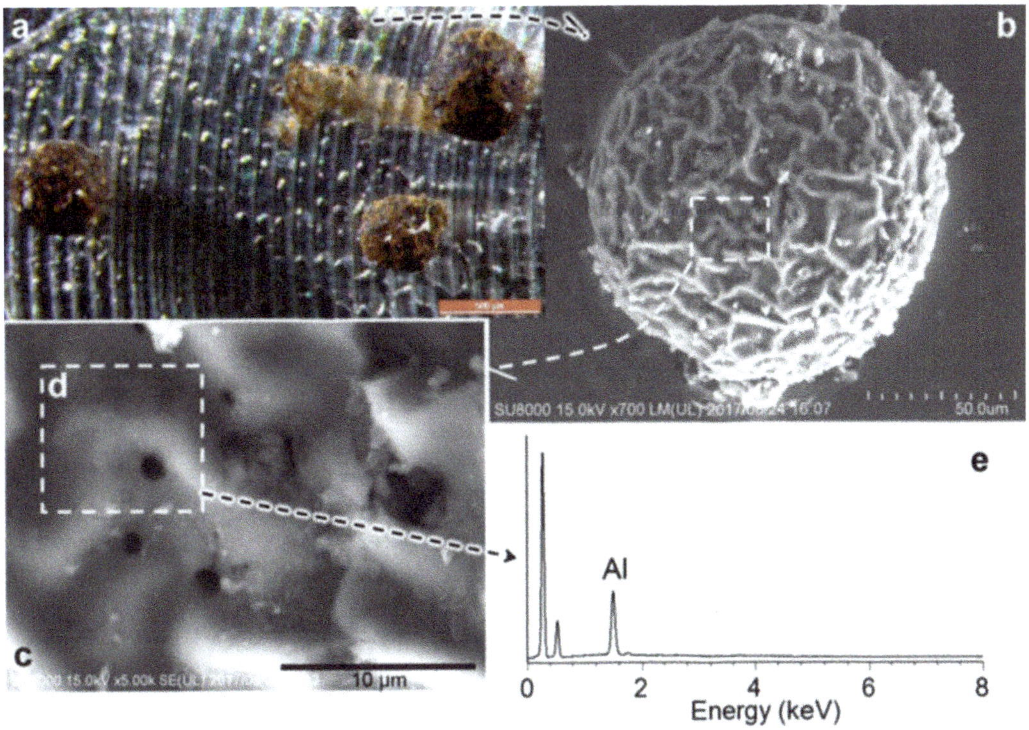

Fig. 4.22 Sclerotia grains separated from the A3 horizon (Fig. 4.19a). (**a**) Optical micrograph of sclerotia grains, (**b**) SEM image of the smallest grain in (**a**), (**c**) magnified SEM image of the dashed area of (**b**), (**e**) EDX spectrum of the dashed area (**d**) in (**c**)

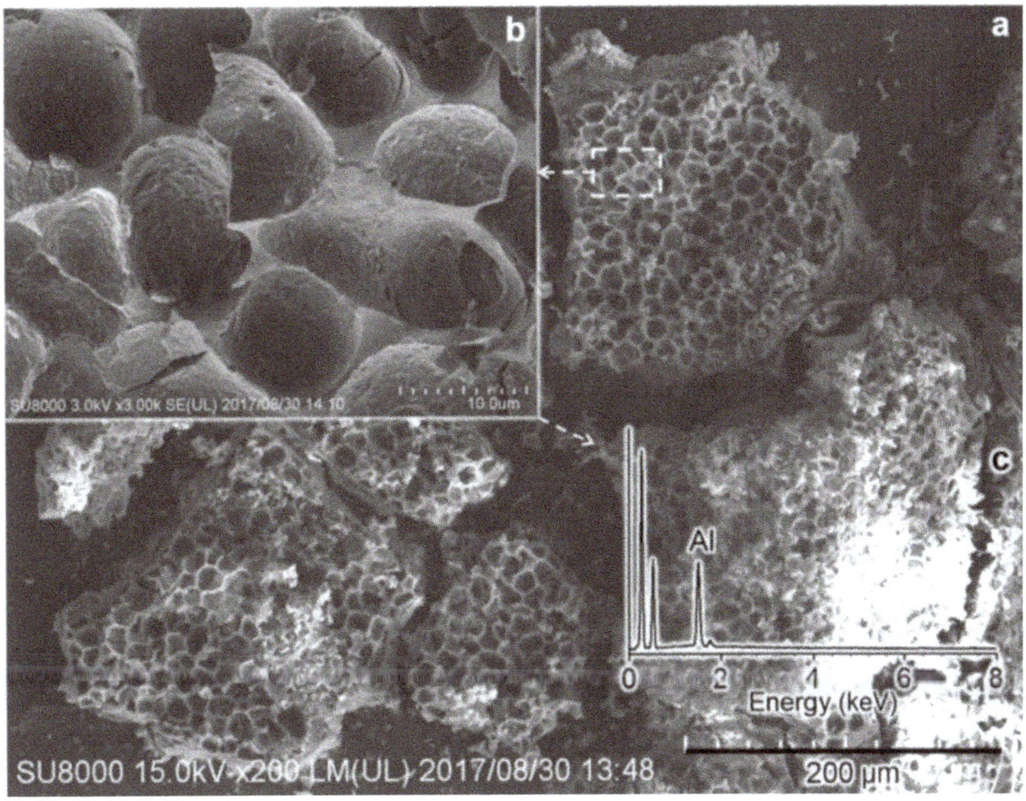

Fig. 4.23 Inside of the largest sclerotia granule shown in Fig. 4.22a. (**a**) SEM image, (**b**) magnified SEM image of the dashed area of (**a**), (**c**) EDX spectrum of (**b**)

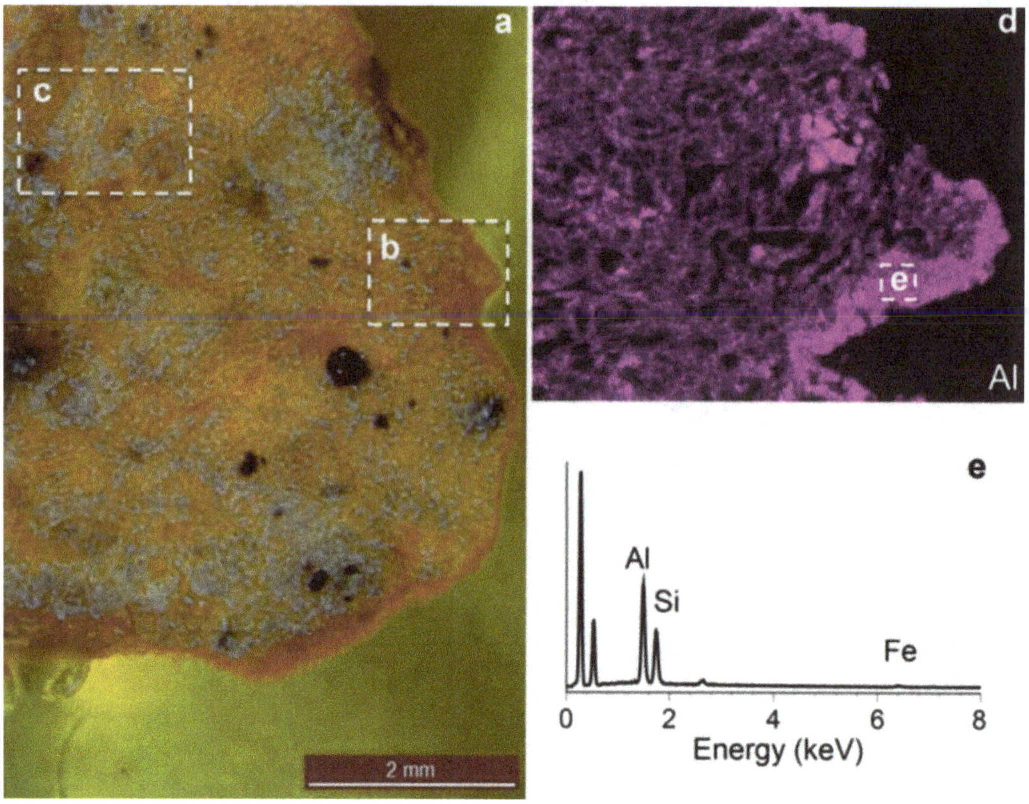

Fig. 4.24 Polished section of a clod from the 2Bw1 horizon shown in Fig. 4.19a. **(a)** Optical micrograph, **(c)** selected area for Fig. 4.25, **(d)** Al element map of the dashed area **(b)** in **(a)**, **(e)** EDX spectrum of the dashed area **(e)** in **(d)**

In Fig. 4.26a, which is the magnified SEM image of dashed square Fig. 4.25d, vesicular structures remain inside the sponge-like volcanic glass. However, two types of vesicle walls can be distinguished: thick and thin. Two thin walls (indicated by red boxes labeled c and d in Fig. 4.26a) were found to be Al-rich according to their EDX spectra (Fig. 4.26c, d). These EDX spectra are close to that for allophanic material. In contrast, a thick wall (indicated by the blue box labelled b in Fig. 4.26a), has an EDX spectrum similar to that of non-colored volcanic glass like Fig. 4.1a–c. Examining the Al and Si element maps of Fig. 4.26a shown in Fig. 4.26e, f, respectively, thin walls in the Fig. 4.26a are Al-rich and Si-poor, whereas thick walls are Al-poor and Si-rich. These observations suggest that the thin walls of vesicles are allophanic altermorphs described by Stoops (2007) and Gerard et al. (2007). Comparing the element maps of Al and Si, Fig. 4.26e, f, respectively, it can be seen that allophanic altermorphs also surround the thick volcanic glass. Hence, the allophanic altermorph formed by releasing Si, Ca, Na, and other soluble elements from the surface of volcanic glass, and remains at the original site.

Figure 4.25e shows a phenocryst that displays cleavage. Examining the cleavage reveals (Fig. 4.27a) thin stripes along the edge of the remaining phenocryst. Examining EDX spectra of the thin stripe (dashed square (b) in Fig. 4.27a) and phenocryst (dashed square (c) in Fig. 4.27a) reveals that they are allophanic material and

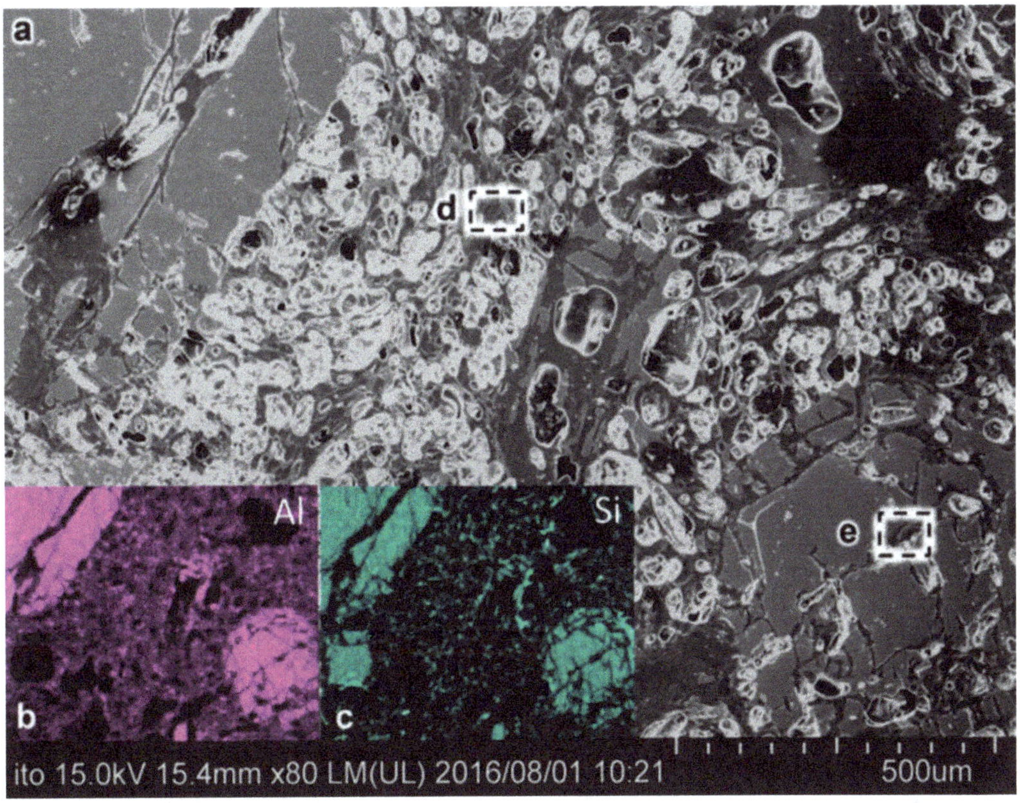

Fig. 4.25 SEM-EDX analyses of the selected area (c) in Fig. 4.24a. (**a**) SEM image, (**b** and **c**) Al, and Si maps of (**a**), respectively

plagioclase, respectively. Examining Al (Fig. 4.27d) and Si (Fig. 4.27e) element maps, thin stripes of the plagioclase phenocrysts are evident. The Al concentration is higher than the Si concentration in these thin stripes as shown by the EDX spectrum (Fig. 4.27b). Two possibilities exist for the formation of the allophanic thin stripes. One is the formation of altermorphs of plagioclase and the other is precipitation of allophanic materials in the cleavage opening.

The 3Bw5 horizon of Fig. 4.19a is a highly weathered example formed from Na-I tephra. The weathered particle is almost wholly reddish yellow to yellow (Fig. 4.28a). The many small vacant bubbles visible in the SEM image of Fig. 4.28b are probably due to inadequate penetration of resin into the small vesicular pores. The Al (Fig. 4.28c) and Si (Fig. 4.28d) element maps show that the majority is Al-rich and Si-poor, except for several small spots in the Si element map. To examine the fine structure of the inside of the particle, a small red square (Fig. 4.28b) was selected and magnified in Fig. 4.29.

The magnified SEM image (Fig. 4.29a) shows that an altermorph of the vesicle walls remains and that materials with diffuse boundaries are filling part of the pores. The EDX spectra of the red squares (b) and (c) in Fig. 4.29a show that these are both Al-rich materials, possibly allophanic in nature. The Al (Fig. 4.29d) and Si (Fig. 4.29e) element maps show that no glassy parts remain. Although the peaks are small, the iron content is higher than that in Fig. 4.26c, d. The reddish yellow

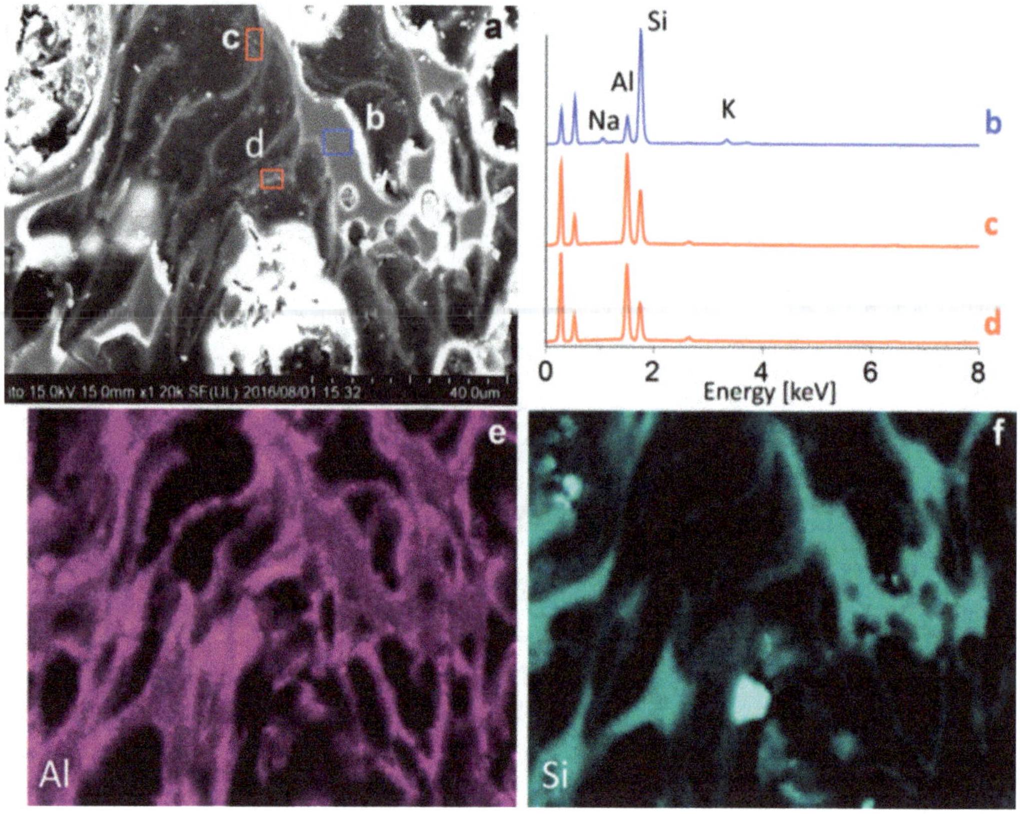

Fig. 4.26 Weathering of volcanic glass inside the Nt-S pumice. **(a)** Magnified SEM image of the selected area Fig. 4.25d, (**b**, **c**, and **d**) EDX spectra of the selected areas (b), (c) and (d) of (a), respectively, (**e** and **f**) Al and Si element maps of (**a**), respectively

color of Fig. 4.28a and the 3Bw5 horizon of Fig. 4.19a is probably owing to iron or ferrihydrite in the allophanic aggregate. The oxalate-extractable Fe (Fe_o) content of the 3Bw5 horizon is 4.78%, and is even higher than the 2Bw1 horizon (Fig. 4.19a, 0.46%). The vesicle walls of the Fig. 4.29a are thin compared to the glass walls remaining in the Fig. 4.26a. Weathering is more intensive in Nt-I than in Nt-S even though the difference in age of the Nt-I and Nt-S is estimated to be small. Possible reasons are that (i) the glass wall was thin, and (ii) the iron content of the volcanic glass was high. Another difference between the 3Bw5 horizon and the scoria particles in Figs. 4.6 and 4.7 is the lack of phenocrysts in the former.

4.4.5 Changes in Elemental Composition with Andisol Formation

The changes in element concentrations are large during the process of Andisol formation (Fig. 4.14) as shown by the differences in Si:Al atomic ratio (Table 4.2) between the parent materials and weathering products. In short, Al-rich products form from Si-rich parent materials. Changes in the concentrations of 57 elements

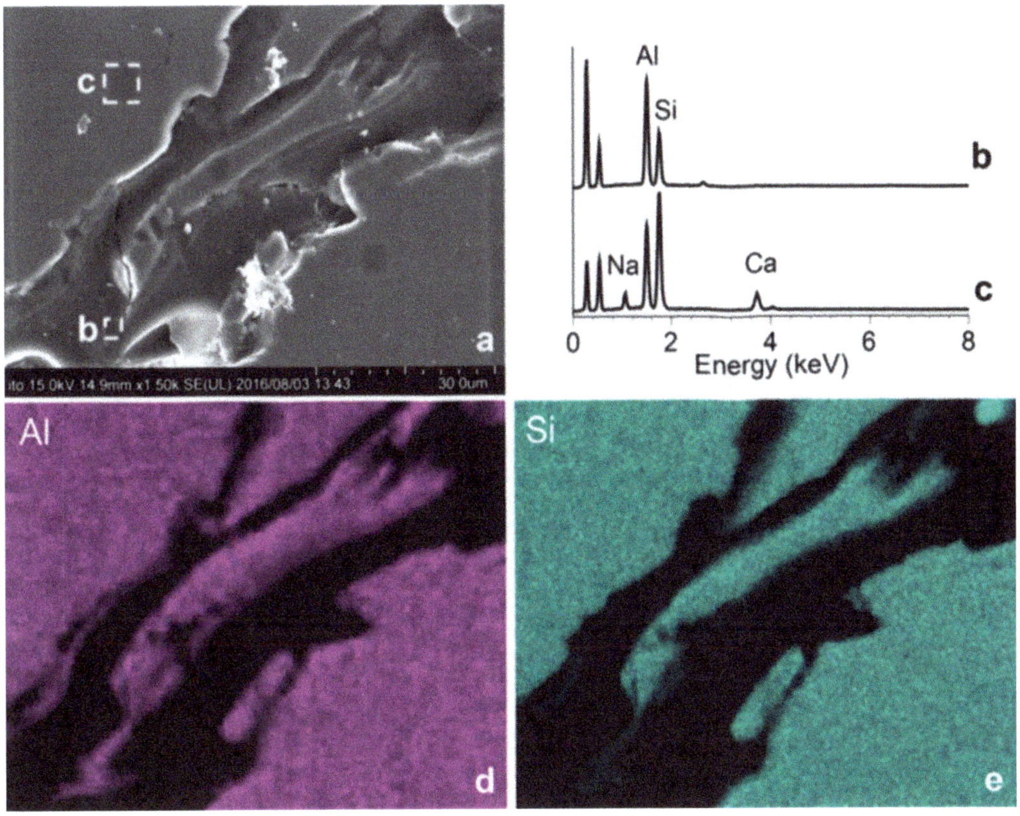

Fig. 4.27 Allophanic materials in the cleavage opening of a phenocryst. (**a**) SEM image of the selected area in Fig. 4.25e, (**d** and **e**) Al and Si element maps of (**a**)

during Andisol formation were reported as a function of Al_o and Si_o (Nanzyo et al. 2002). To discuss the mechanism of the changes in element concentration during the process of Andisol formation, we constructed Fig. 4.30 (Nanzyo et al. 2007). Horizontal axes show an index of the weight ratio $(W_p - W_s)/W_s = W_p/W_s - 1$, where W_p is the weight of parent volcanic ash, and W_s the weight of Andisol. Vertical axes show the concentrations of the elements, although the absolute concentration range depends on the individual element. The element concentration (E_j) at $W_p/W_s - 1 = 0$, with rearrangement $W_p = W_s$, is the element concentration present in the parent volcanic ash ($E_{j,p}$). The element concentration at $W_p/W_s - 1 = 2$, which is the same as $W_p = 3W_s$, is the element concentration when the weight of the Andisol has been reduced to one-third of the weight of the parent volcanic ash. If an element j is immobile during Andisol formation, Ej at $W_p/W_s - 1 = 2$ is $3E_{j,p}$.

Two analyses were performed before constructing Fig. 4.30. One was a principal component analysis of element concentration data of 46 elements using 95 Andisol samples from 18 pedons with different rock types. The first principal component was suggested to be depletion and enrichment of elements, and the second principal component (PC2) was suggested to be related to rock type. Then, andesitic Andisols, for which the variation in PC2 scores was relatively small, was used for Fig. 4.30.

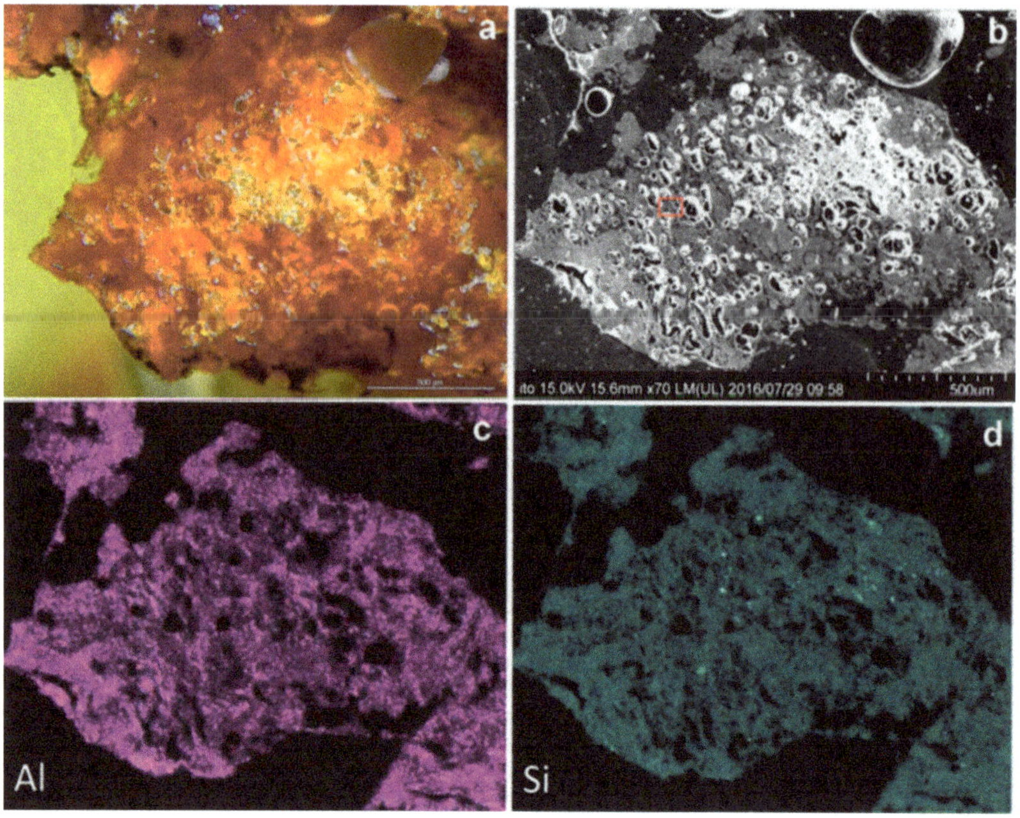

Fig. 4.28 Polished section of a clod from the 3Bw5 horizon of Fig. 4.19a. (**a**) Optical micrograph, (**b**) SEM image, (**c** and **d**) Al and Si element maps, respectively

The other preliminary analysis was to estimate W_p/W_s value for each sample. Assuming that volcanic glass quantitatively weathered to form Al_o materials, and the average Al content of volcanic glass in the andesitic volcanic ash is 69.6 g kg^{-1} (Kobayashi et al. 1976); W_p/W_s was calculated as follows on an ignition residue basis:

$$W_p/W_s = \mathrm{OIIR} + Al_o/69.6 \tag{4.1}$$

where OIIR (kg kg^{-1}) is the oxalate-insoluble residue, and was determined experimentally.

In Fig. 4.30, the ideal concentration changes of immobile elements were also drawn as a red line. The use of the red lines in this figure is based on the open-system mass transport that yields the chemical gain and losses of elements in a soil sample compared with the parent material after Brimhall et al. (1991) and Nieuwenhuyse and van Breemen (1997). For immobile elements, their equation can be simplified by multiplying the volume by the bulk density to give Eq. (4.2) (Kurtz et al. 2000):

$$W_p E_{j,p} = W_s E_{j,s} \tag{4.2}$$

Equation (4.2) is further converted to Eq. (4.3):

$$E_{j,s} = \left(W_p/W_s\right) E_{j,p} \tag{4.3}$$

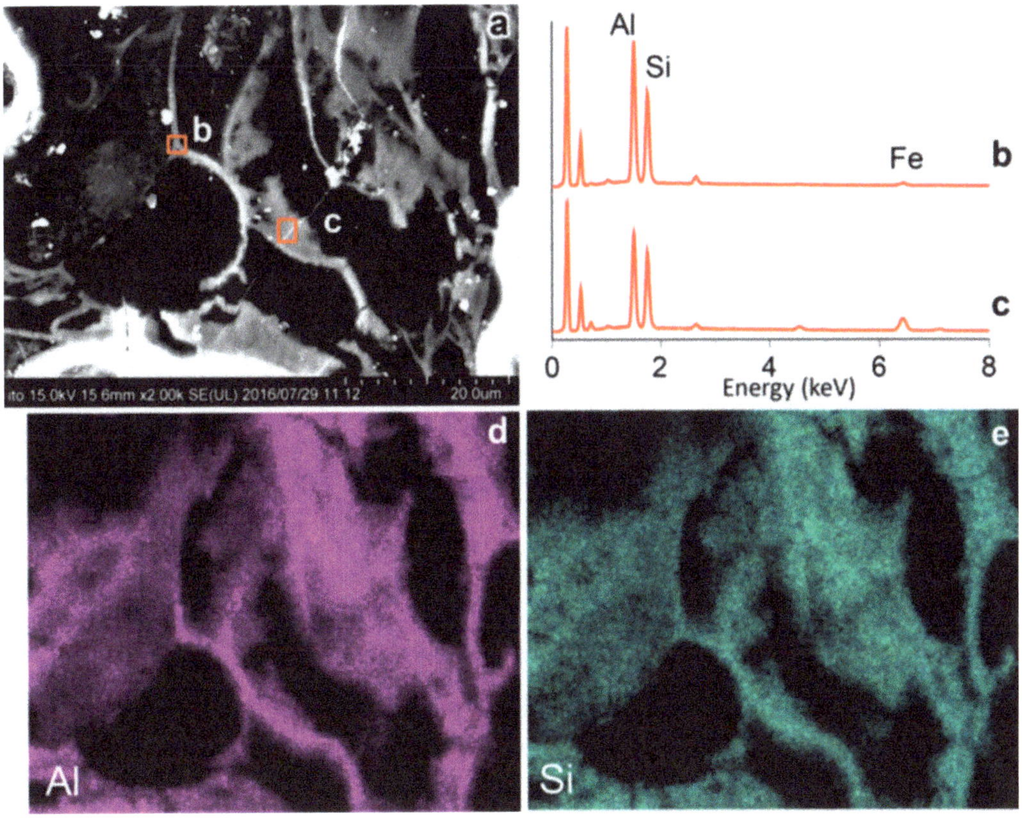

Fig. 4.29 Weathering of volcanic glass inside the Nt-I pumice. (**a**) SEM image of the selected area of Fig. 4.28b, (**b** and **c**) EDX spectra of the selected areas (b) and (c), respectively, (**d** and **e**) Al and Si element maps of (**a**), respectively

As W_p/W_s is calculated using Eq. (4.1), the red lines in Fig. 4.30 can be drawn so as to pass through the average concentration $E_{j,s}$ and the average weight change W_p/W_s. If an element is immobile during Andisol formation, its concentration is plotted along this line. The theoretical $E_{j,p}$ can also be calculated using $E_{j,s}$ and W_p/W_s so long as an element is immobile. If a parent tephra has higher $E_{j,p}$ than average, the plot of $E_{j,s}$ will appear above the solid line, and if it has a lower $E_{j,p}$, it will appear below the solid line. If different $E_{j,p}$ values of elements are scaled similarly, the slopes of their $E_{j,s}$ plots will also be similar as seen in Fig. 4.30.

Among 54 elements, at least 27 (Be, Al, Ti, Fe, Y, Zr, Nb, La, Ce, Pr, Nd, Sm, Eu, Gd, Tb, Dy, Ho, Er, Tm, Yb, Lu, Hf, Ta, Tl, Pb, Th, and U) were enriched in the Andisols, and the increase in these concentrations was related to total weight loss owing to the soil formation processes. Of the major elements in soil, concentrations of Si, Ca, and Na clearly decrease with the weight loss as shown in Fig. 4.30. However, the slope of the decrease in the element concentration is steeper for Ca and Na than for Si (Fig. 4.30). One possible reason is that Ca and Na are not the major constituents of Andisol formation products, but Si is the main structural constituent of allophane and imogolite. Other possible explanations are that Si can be sorbed by ferrihydrite, and that Si is also partly retained in Andisols as opaline silica or phytoliths. The weight-loss of tephra, maintaining the altermorphs (Figs. 4.26 and

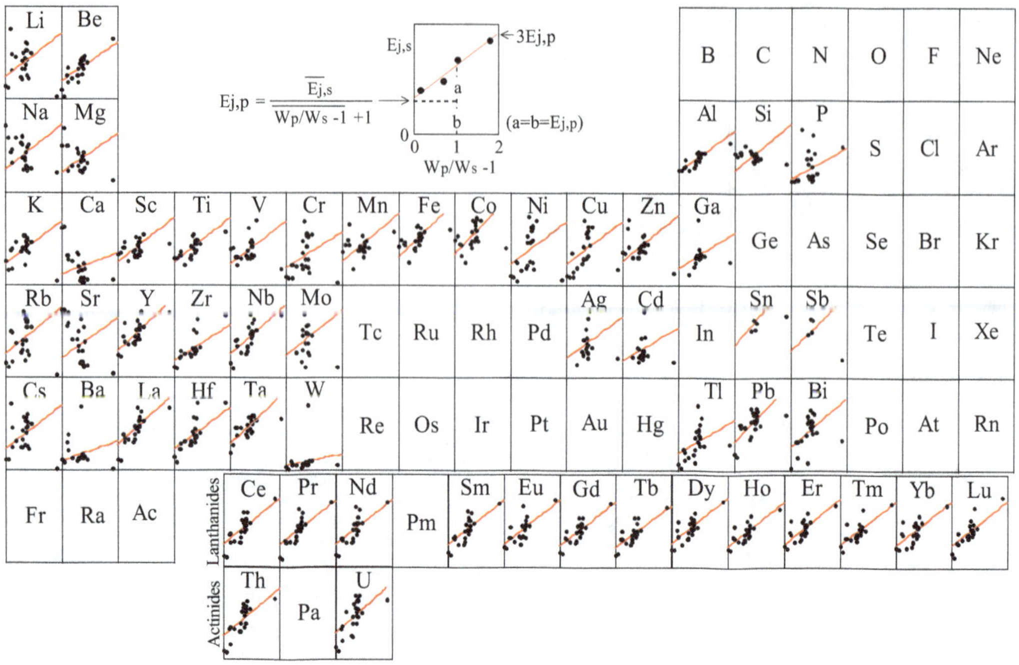

Fig. 4.30 Changes in element concentration of andesitic tephra during Andisol formation. Red lines show changes in concentration of elements with weight loss of soils if the elements are immobile

4.29), may be the major reason for the reduction of bulk density in the Bw horizon of Andisols (Nanzyo et al. 2007).

4.4.6 *Volcanic Ash Soils Under Various Drainage Conditions*

The weathering of volcanic ash depends on moisture and drainage conditions in soil. Under moist and well-drained conditions, active Al and Fe materials form as summarized in Fig. 4.14. In contrast, halloysite formation prevails in volcanic ash soils with restricted drainage or under semi-dry weather conditions. Figure 4.31 shows changes in the weathering of dacitic volcanic ash Numazawa-Numazawako (Nm-NK) (4.6 ka, Ishizaki et al. 2009; Yamamoto 2003, 2007) with topography and drainage conditions at Aizu basin, Japan. In this area, tephra from the Numazawa caldera is distributed on the buried A horizon soil. At the well-drained hilly sites (Jotohara and Urushihara in Fig. 4.31a, b), the color of Nm-NK tephra is brownish, the range of $Al_o + Fe_o/2$ values is 12–19 g kg^{-1}, and the clay fractions are mainly composed of allophane and imogolite with small amounts of halloysite. On the other hand, at the Yukawa and Amanuma sites (Fig. 4.31d, e) with restricted drainage, the color of the Nm-NK is grayish, the range of $Al_o + Fe_o/2$ values is 1.2–5.3 g kg^{-1}, and the major clay mineral is halloysite. Although the Nm-NK tephra is thick at the Yukawa site, the reason may be secondary deposition from sounding areas due to transportation by water. At Sobanome (Figs. 3.6a, 4.31c), which is also inside the

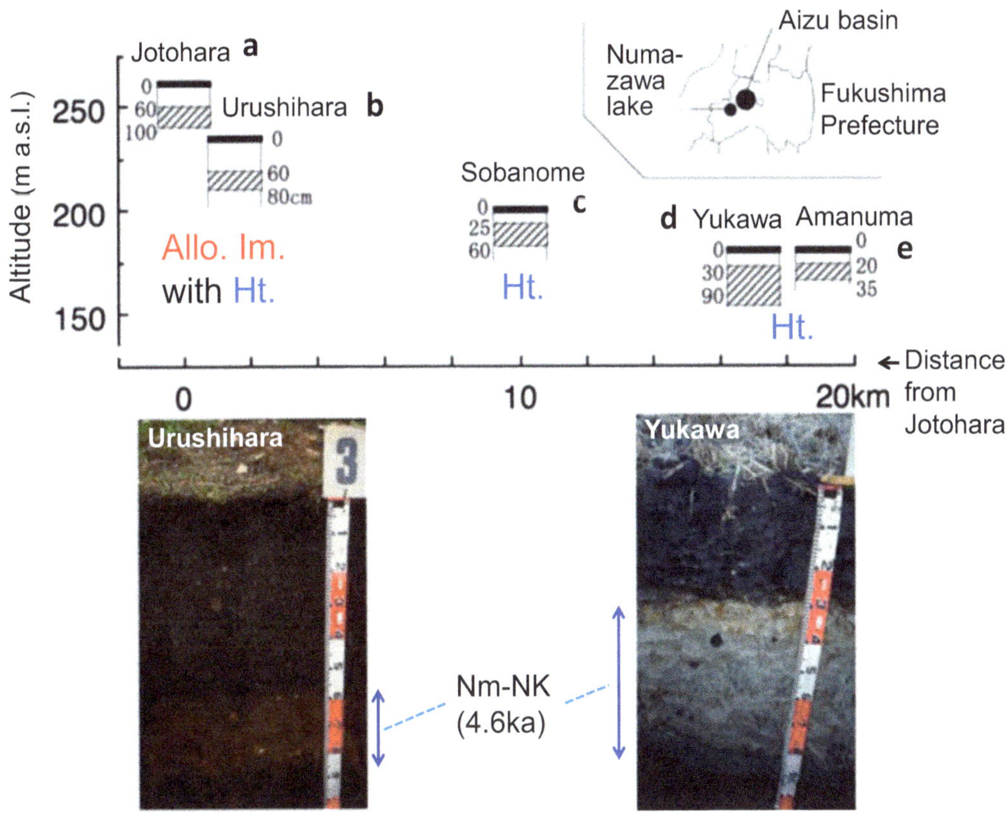

Fig. 4.31 Major weathering products, and examples of soil profile including Nm-NK tephra in Aizu, Japan. Jotohara is the nearest study site to Numazawa caldera lake (**a, b**: good drainage, **c, d, e**: restricted drainage). The shaded position of (**a**) to (**e**) shows Nm-NK tephra at each site, and major weathering products "Allo.", "Im." and "Ht." denote allophane, imogolite, and halloysite, respectively

basin and poorly drained, the color of the Nm-NK is orange-gray, the range of $Al_o + Fe_o/2$ values is 4.8–6.6 g kg^{-1}, and the major clay mineral is halloysite. Changes in weathering of volcanic glass with drainage conditions were also observed by Stoops (2007).

References

Aramaki S (1984) Formation of the Aira Caldera, southern Kyushu, –22,000 years ago. J Geophys Res 89:8485–8501

Arnalds O, Bartoli F, Buurman P, Oskarsson H, Stoops G, Garcia-Rodeja E (2007) Soils of volcanic regions in Europe. Springer, Berlin Heiderberg

Baer EM, Fisher RV, Fuller M, Valentine G (1997) Turbulent transport and deposition of the Ito pyroclastic flow: determination using anisotropy of magnetic susueptivility. J Geophys Res 102:22565–22586

Brimhall GH, Chadwick OA, Lewis CJ, Compston W, Williams IS, Danti KJ, Dietrich WE, Power ME, Hendricks D, Bratt J (1991) Deformational mass transport and invasive processes in soil evolution. Science 255:695–702

Busacca AJ, Marks HM, Rossi R (2001) Volcanic glass in soils of the Columbia Plateau, Pacific Northwest, USA. Soil Sci Soc Am J 65:161–168

Cuadros J, Cabalero E, Huertas FJ, de Cisneros CJ, Huertas F, Linares J (1999) Experimental alteration of volcanic tuff: smectite formation and effect of 18O isotope composition. Clay Clay Min 47:769–776

Dahlgren RA (1994) Quantification of allophane and imogolite. In: Amonette JE, Zelazny LW (eds) Quantitative methods in soil mineralogy. SSSA Miscellaneous Publication, SSSA, Madison, pp 430–451

Dahlgren RA, Shoji S, Nanzyo M (1993) Mineralogical characteristics of volcanic ash soils. In: Shoji S, Nanzyo M, Dahlgren RA (eds) Volcanic ash soils –genesis, properties and utilization. Elsevier, Amsterdam, pp 101–143

Drees LR, Wilding LP, Smeck NE, Senkayi AL (1989) Silica in soils: quartz and disordered silica polymorphs. In: Dixon JB, Weed SB (eds) Minerals in soil environments, 2nd edn. SSSA, Madison, pp 913–974

Eden DN (1992) A standard method for determining volcanic glass content in Andisols. DSIR Land Resources Scientific Report No.2. DSIR Land Resources, Private Bag, Lower Hutt, New Zealand, pp 5–27

Gerard M, Caquieau S, Pinheiro J, Stoops G (2007) Weathering and allophane neoformation in soils developed on volcanic ash in the Azores. Eur J Soil Sci 58:496–515

Harsh J, Chorover J, Nizeyimana E (2002) Allophane and imogolite. In: Dixon JB, Schulze DG (eds) Soil mineralogy with environmental applications. SSSA Inc., Madison, pp 291–322

Heiken G, Wohletz K (1985) Volcanic ash. University of California Press, Berkeley, pp 1–246

Henmi T, Parfitt RL (1980) Laminar opaline silica from volcanic ash soils in New Zealand. Clay Clay Miner 28:57–60

Ishizaki Y, Masubuchi Y, Ando Y (2009) Two types of dacite pumices from the caldera-forming eruption of Numazawa Volcano, NE Japan. J Miner Petrol Sci 104:356–373

Ishizaki Y, Morita T, Toriyama H (2017) Sequence and magma plumbing system of explosive eruptions that formed the Nantai-Imaichi tephra and the associated scoria flow deposits, Nantai volcano, NE Japan. Bull Volc Soc Jpn 62:95–116

Ito T, Shoji S, Shirato Y, Ono E (1991) Differentiation of a spodic horizon from a buried A horizon. Soil Sci Soc Am J 55:438–442

Kawai K (1969) Micromorphological studies of Andosols in Japan. Bull Ntnl Inst Agric Sci (Japan) 20:77–154

Kobayashi S, Shoji S, Yamada I, Masui J (1976) Chemical and mineralogical studies on volcanic ashes, III. Some mineralogical and chemical properties of volcanic glasses with special reference to the rock types of volcanic ashes. Soil Sci Plant Nutr 22:7–13

Kobayashi J, Takada R, Nakano S (2007) Eruptive history of Fuji Volcano from AD 700 to AD 1000 using stratigraphic correlation of the Kozushima-Tenjosan tephra. Bull Geol Sur Jpn 57:409–430

Kondo R, Sase T, Kato Y (1988) Opal phytolith analysis of andisols with regard to interpretation of paleovegetation. In: Kinloch DI, Shoji S, Beinroth FH, Eswaran E (eds) Proceedings of the 9th international soil classification workshop, Japan. 20 July to 1 August, 1987. Publication by Japan Committee of 9th international soil classification workshop, for the soil management support services, Washington, DC, USA, pp 520–534

Kumada K, Hurst HM (1967) Green humic acid and its possible origin as a fungal metabolite. Nature 214:631–633

Kurtz AC, Derry LA, Jo Alfano M, Chadwick OA (2000) Refractory element mobility in volcanic soils. Geology 28:683–686

Le Maitre RW (2002) Igneous rocks: a classification and glossary of terms, recommendations of the international union of geological sciences, subcommission of the systematicsl of igneous rocks. Cambridge University Press, Cambridge

Lynn W, Thomas JE, Moody LE (2008) Petrographic microscope techniques for identifying soil minerals in grain mounts. In: Ulery AL, Drees LR (eds) Methods of soil analysis. Part 5. Mineralogical methods, SSSA book series, vol 5. SSSA, Madison, pp 161–190

Machida H (2002a) Quaternary volcanoes and widespread tephras. Glob Environ Res 6:3–17

Machida H (2002b) Volcanoes and tephras in the Japan area. Glob Environ Res 6:19–28

McDaniel PA, Lowe DJ, Arnalds O, Ping C-L (2012) Andisols. In: Huang PM, Li Y, Sumner ME (eds) Handbook of soil sciences, properties and processes, vol 33, 2nd edn. CRC Press, Taylor & Francis Group, Boca Raton-London-New York, pp 29–48

Miyaji N (2002) The 1707 eruption of Fuji volcano and its tephra. Glob Environ Res 6:37–39

Mizuno N, Amano Y, Mizuno T, Nanzyo M (2008) Changes in the heavy minerals content of Tarumae-a tephra with distance from the source volcano and its effect on the element concentration of the tephra. Soil Sci Plant Nutr 54:839–845

Nanzyo M, Yamasaki S, Honna T (2002) Changes in content of trace and ultratrace elements with an increase in noncristalline materials in volcanic ash soils of Japan. Clay Sci 12:25–32

Nanzyo M, Yamasaki S, Honna T, Yamada I, Shoji S, Takahashi T (2007) Changes in element concentrations during Andosol formation on tephra in Japan. Eur J Soil Sci 58:465–477

Nieuwenhuyse A, van Breemen N (1997) Quantitative aspects of weathering and neoformation in selected Costa Rican volcanic soils. Soil Sci Soc Am J 61:1450–1458

Otowa M, Shoji S (1987) Ninth international soil classification workshop, properties, classification, and utilization of Andiosls and Paddy soils, Kanto, Tohoku, and Hokkaido, Japan, 20 July–1 August 1987, Tour Guide. Soil Management Support Survice, USDA Soil Concervation Service, Washington DC, Japanese Committee of the ninth international soil classification workshop, pp 241–251

Parfitt RL, Henmi T (1982) Comparison of an oxalate extraction method and an infrared spectroscopic method for determining allphane in soil clays. Soil Sci Plant Nutr 28:183–190

Parfitt RL, Wilson AD (1985) Estimation of allophane and halloysite in three sequences of volcanic soil, New Zealand. In: Caldas EF, Yaalon DH (eds) Volcanic soils, weathering and landscape relationships of soils on tephra and basalt, catena supplement 7. Catena Verlag, Braunschweig, pp 1–8

Shoji S (1986) Mineralogical characteristics I. Primary minerals. In: Wada K (ed) Andosols in Japan. Kyushu University Press, Fukuoka, pp 21–40

Shoji S, Masui J (1969) Amorphous clay minerals of recent volcanic ash soils in Hokkaido (I). Soil Sci Plant Nutr 15:191–201

Shoji S, Masui J (1971) Opaline silica of recent volcanic ash soils in Japan. J Soil Sci 22:101–112

Shoji S, Saigusa M (1978) Occurrence of laminar opaline silica in some Oregon andosols. Soil Sci Plant Nutr 24:157–160

Shoji S, Kobayashi S, Yamada I, Masui J (1975) Chemical and mineralogical studies o volcanic ashes. I. Chemical composition of volcanic ashes and their classification. Soil Sci Plant Nutr 21:311–318

Shoji S, Ito T, Saigusa M, Yamada I (1985) Properties of nonallophanic Andosols from Japan. Soil Sci 140:264–277

Shoji S, Nanzyo M, Dahlgren RA (1993) Volcanic ash soils – genesis, properties and utilization. Elsevier, Amsterdam

Sievert L, Simkin T, Kimberly P (2010) Volcanoes of the world. University of California Press, Berkeley/Los Angeles, pp 48–49

Sonehara T (2016) Chemical composition of volcanic glass samples determined by EPMA and XRF: examples of the Ito and Aso-4 pyroclastic flow deposits. Eng Geol J 6:5–19 (in Japanese)

Stoops G (2007) Micromorphology of soils derived from volcanic ash in Europe: a review and synthesis. Eur J Soil Sci 58:356–377

Takahashi T, Dahlgren RA (2016) Nature, properties and function of aluminum-humus complexes in volcanic soils. Geoderma 263:110–121

Tomita K, Yamane H, Kawano M (1993) Synthesis of smectite from volcanic glass at low temperature. Clays Clay Miner 41:655–661

Ugolini FC, Dahlgren R, Shoji S, Ito T (1988) An example of andosolization as revealed by soil solution studies, southern Hakkoda, northeastern Japan. Soil Sci 145:111–125

Wada K (1989) Allophane and imogolite. In: Dixon JB, Weed SB (eds) Minerals in soil environments, 2nd edn. SSSA, Madison, pp 1051–1087

Wada S-I, Nagasato A (1983) Formation of silica microplate by freezing dilute silicic acid solution. Soil Sci Plant Nutr 29:93–95

Watanabe M, Fujitake N, Ohta H, Yokoyama T (2001) Aluminum concentrations in sclerotia from a buried humic horizon of volcanic ash soils in Mt. Myoko, Central Japan. Soil Sci Plant Nutr. 47:411–418

Watanabe M, Kato T, Ohta H, Fujitake N (2002) Distribution and development of sclerotium grain as influenced by aluminum status in volcanic ash soils. Soil Sci Plant Nutr 48:569–575

Watanabe M, Genseki A, Sakagami N, Inoue Y, Ohta H, Fujitake N (2004) Aluminum oxyhydroxide polymorphs and some micromorphological characteristics in sclerotuim grains. Soil Sci Plant Nutr 50:1205–1210

Watanabe M, Inoue Y, Sakagami N, Bolormaa O, Kawasaki K, Hiradate S, Fujitake N, Ohta H (2007) Characterization of major and trace elements in sclerotium grains. Eur J Soil Sci 58(3):786–793

Wilding LP, Smeck NE, Drees LR (1977) Silica in soils: quartz, cristobalite, tridymite and opal. In: Dixon JB, Weed SB, Kittrick JA, Milford MH, White JL (eds) Minerals in soil environments. SSSA, Madison, pp 471–552

Yamada I, Shoji S (1975) Relationship between particle-size and mineral composition of volcanic ashes. Tohoku J Agr Res 26:7–10

Yamamoto T (2003) Eruptive history of Numazawa volcano, NE Japan: new study of the stratigrahy, eruption ages, and eruption volumes of the products. Bull Geol Surv Japan 54:323–340 (in Japanese with English abstract)

Yamamoto T (2007) A rhyolite to dacite sequence of volcanisim directly from the heated lower crust: Late Pleistocene to Holocene Numazawa volcano, NE Japan. J Volcanol Geotherm Res 167:119–133

Yonebayashi K, Hattori T (1988) Chemical and biological studies on environmental humic acids, I. Composition of elemental and functional groups of humic acid. Soil Sci Plant Nutr 34:571–584

Yoshinaga N (1986) Mineralogical characteristics II. Clay minerals. In: Wada K (ed) Andosols in Japan. Kyushu University Press, Fukuoka, pp 41–56

Chapter 5
Inorganic Soil Constituents Sensitive to Varying Redox Conditions

Abstract Inorganic soil constituents sensitive to varying redox conditions, such as hydrated iron oxide, vivianite, siderite, iron (II) sulfides, and jarosite, are analyzed using optical and electron microscopes, energy dispersive X-ray spectroscopy (EDX), and X-ray diffraction (XRD). Many of these minerals are sourced from paddy field soils, which undergo reducing and oxidizing conditions in the plow layer every year. Iron mottles formed at the soil redox interface in the presence of reducing and oxidizing conditions provide significant visual evidence of varying redox conditions in soil. Polished sections were used to examine the elemental distributions and morphological properties of the mottles. One type of iron mottles is formed around rice roots by oxygen diffusion from the roots. They are cylindrical in form and include soil matrix minerals. Other type of iron mottles is formed on the surfaces of irregular or vesicular pores by oxygen diffusion through soil pores after drainage. These mottles contain few soil matrix minerals. In association with iron, the distribution of phosphate is strongly affected by changes in redox conditions in paddy field soils with low active Al content.

5.1 Introduction

Oxidizing conditions are common in surface soils in an oxygen-containing atmosphere. Under oxidizing conditions, the major factors affecting dissolution and precipitation reactions in soil are the pH, temperature, and elemental composition of the soil. For example, Ca carbonate dissolves in acid soils, whereas it precipitates in alkaline soils. However, under submergence or high ground water level, reducing conditions develop in soil associated with microbial activity. For example, the solubility of iron is very low in neutral and alkaline soils, whereas a part of iron dissolves under reducing conditions. Thus, changes in redox conditions affect the chemical forms of elements, including redox-sensitive elements and related elements (Ponnamperuma 1972). The morphological properties of soil profiles are also affected by these redox reactions (Vepraskas and Craft 2016).

Reduced soils are typically characterized by the reduction of iron, as included in the diagnostic criteria for redoximorphic features in the United States Department of

M. Nanzyo, H. Kanno, *Inorganic Constituents in Soil*,
https://doi.org/10.1007/978-981-13-1214-4_5

Agriculture – Soil Taxonomy (USDA-ST) (Soil Survey Staff 1999) and gleyic properties in the World Reference Base for Soil Resources (WRB) (IUSS Working Group WRB 2015). Iron is one of the major elements in soil, and it strongly affects soil color and mottle formation.

In relation to redox reactions, the soil pH also changes. For example, pH increases with reduction of hydrated iron (III) oxide, and decreases with oxidation of ferrous iron (iron(II)). A pH–pE diagram can be used as a method to describe the chemical stability of minerals under varying redox conditions in soil, and is recommended for further study (Stumm and Morgan 1996; Kyuma 2004).

5.1.1 Alternating Oxidized and Reducing Conditions in Paddy Field Soils

Paddy field soils are an example of soil with varying redox conditions. The total paddy-field land area comprises irrigated, rain-fed lowland, and rain-fed upland paddy fields covering 93, 52, and 15 million ha (Global Rice Science Partnership 2013), respectively. More than 90% of the paddy fields in the world are distributed in Asia. Elsewhere, paddy fields are distributed in temperate, subtropical, and tropical areas where enough water is available. The rain-fed lowland paddy field area includes that covered by deep-water rice. Although significant areas of irrigated paddy fields are also grown in rotation with a range of other crops, lowland paddy field soils experience relatively reducing conditions when rice is grown under submergence. Merits of submergence are (i) high rice yield, (ii) weed control, (iii) an increase in phosphate availability under reducing conditions, (iv) supply of micronutrients as solutes in irrigation water, and (v) high N-fixation ability compared to ordinary uplands (Kyuma 2004). This chapter focuses on paddy field soils, and lowland soils with high ground water level in relation to their varying redox conditions.

Reducing conditions are caused by microbial activity. Three typical requirements for the development of reducing conditions are (i) submergence of soil in water to restrict oxygen diffusion, (ii) appropriate temperature for microbial activity, and (iii) carbon source for microbes. Oxidizing forms of C, N, Mn, Fe, and S are also important, and nearly-neutral soil pH is preferable for microbial activity. In contrast, when reduced soil is exposed to air due to drainage or ground water level fall, the soil becomes oxidized.

Major elements affected by varying redox conditions in soil are C, N, Mn, Fe, and S. The behavior of contaminant elements in soil, such as Cd, Cu, and As, is also affected by varying redox conditions. Sulfide precipitation of Cd and some other heavy metals is possible under reducing conditions. The oxidation numbers of Cu and As change with redox potential.

Figure 5.1 illustrates changes in the color of soil and rice roots as a result of redox reactions. An Ap horizon (plow layer) soil (Udifluvent, according to the USDA-ST), which was sampled from a paddy field, was used. The oxalate-extractable Fe (Fe_o),

Fig. 5.1 Changes in soil color with submergence. (**a**) 4 days after submergence, mixing, and rice transplanting (June 1, 2012), (**b**) 31 days after submergence (June 28, 2012) in a glass vessel

oxalate-extractable Al (Al_o), total organic carbon (TOC), and cation exchange capacity (CEC) values are 8.6, 1.2, 23.0 g kg^{-1}, and 23.6 $cmol_c$ kg^{-1}, respectively (Kusunoki et al. 2015). The major iron mineral in the paddy field soil appears to be poorly crystalline ferrihydrite (Childs et al. 1991; Hansel et al. 2001; Fu et al. 2016). These chemical properties are common to the plow layer of lowland paddy field soils in Japan. Figure 5.1a shows the original brown soil color, and the roots of transplanted rice are white and short.

At 31 days after transplanting, the soil color turned grayish due to reduction of ferrihydrite to ferrous iron (Fig. 5.1b). The largest portion of the ferrous iron appears to remain as exchangeable Fe^{2+} in the soil. At the boundary between the water and the reduced soil, brownish soil, which is called an oxidative layer, remains due to diffusion of oxygen from the air. There are very small dark-colored areas between the oxidative layer and the underlying reduced soil. This dark color resembles noncrystalline ferrous sulfide, as shown in Sect. 5.5.1.

The number of rice roots increased in Fig. 5.1b, and the roots are whitish and brown in color. The rice roots can be classified into three groups based on their diameter: (i) thick (0.5–1 mm), (ii) intermediate (approximately 0.3 mm), and (iii) thin (0.1–0.15 mm). The intermediate and thin roots develop around the thick roots. The very young roots are whitish. Associated with the formation of aerenchyma and lysigenous intercellular space (Kawai et al. 1998), the color of both the thin and thick roots turns brown. This brown color is due to diffusion and concentration of ferrous iron, followed by oxidation of the ferrous iron by oxygen diffusion through the rice roots (Ando et al. 1983; Sadana and Claassen 1996), and precipitation of ferric iron (iron(III)), resulting in the formation of iron plaque (Kahn et al. 2016). Thus, the main redox interface sites in paddy field soil are located between the oxidative soil layer and underlying reduced soil, and between the rice roots and reduced soil. The other redox interface in paddy field soil is between the reduced soil and the underlying oxidative subsoil. Reducing conditions can be found most clearly using the dipyridyl test, which detects the presence of Fe^{2+}.

The color change of a soil profile with an increase in reducing conditions may depend on soil properties. Since the color of the Ap horizon in Andisols is dark, where the dark color is due to high content of highly-humified humus, the color change with an increase in reducing conditions, such as in Fig. 5.1, may be masked even when Fe^{2+} is detected with the dipyridyl test. Fading of the light brown color of goethite aggregates in an incubated soil under submergence is slow, possibly due to the high stability of goethite crystals.

5.1.2 Redox Reactions in Soil

Important redox reactions in the plow layer of a paddy field soil under submergence are shown schematically in Fig. 5.2, where the background figure is similar to Fig. 5.1b. With a decrease in the soil Eh value (The potential that is generated between an oxidation or reduction half-reaction and the standard hydrogen electrode (0.0 V at pH = 0), (Glossary of Soil Science Committee 2008)), several or more reactions proceed in order of decreasing Eh^0(pH 7), as shown in Fig. 5.2c. The reason for using Eh^0(pH 7) in Fig. 5.2c is that the pH(H_2O) value of the paddy field Ap horizon soil after reduction is around 7, whereas the pH of the same air-dried soil

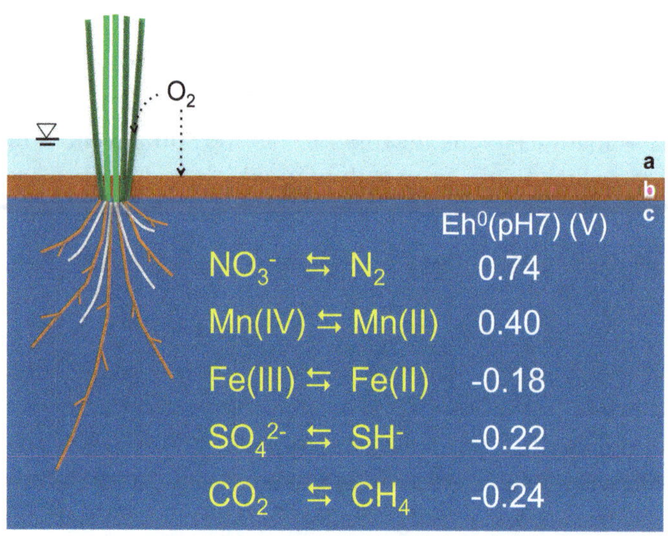

Fig. 5.2 Simplified redox reactions in the plow layer soil of a paddy field under submergence. (**a**) irrigation water, (**b**) oxidizing layer, (**c**) reducing layer). "Fe(III)" denotes $Fe(OH)_3$ (amorphous). For more information, see Stumm and Morgan (1996), Kyuma (2004)

typically ranges between 5 and 6. The pH change between oxidized and reduced soils is exemplified by the reaction $Fe(OH)_3$ (amorphous) ⇄ Fe^{2+} (negatively charged site) + $3OH^-$ + e^-.

The Eh value of irrigation water (Fig. 5.2a) is high due to the addition of O_2 from the air. In the oxidizing layer (Fig. 5.2b), the Eh value decreases steeply to around −0.2 V at the boundary between the oxidizing layer (Fig. 5.2b) and the reducing layer (Fig. 5.2c), estimated from the presence of dark-colored noncrystalline iron sulfide (Fig. 5.1b). The lowest Eh value in the reducing layer is typically −0.2 ~ −0.3 V. The thickness of the oxidizing layer ranges between around 0.5 and 5 cm.

Oxygen diffuses from the air to the rice roots through the aerenchyma, whereas the bulk soil is reduced under submergence. A redox interface is also formed between the aerenchyma and the bulk soil. At this redox interface, there are three or more cell layers, which are the epidermis, exodermis, and sclerenchyma (one or more layers), forming the outer part of a rice root (Kondo et al. 2000). With aging, the epidermis layer is sloughed off first, whereas the exodermis and/or sclerenchyma remain along with deposition of hydrated iron oxide and other materials.

Methane can be formed by reduction of CO_2, which may form through oxidation of organic matter by microbes. Carbon dioxide and CH_4 gases can form vesicular or irregular pores in the submerged and reduced soil. Although the vapor phase ratio of the puddled plow layer soil is only 1–3%, it increases to 9–12% at around 40 days after puddling and submergence, possibly due to formation of these gases (Saito and Kawaguchi 1971a, b). One can easily notice the existence of gases from bubbling when he steps in the submerged paddy field 1 or 2 months after submergence. Methane formed in the reduced soil can be released to the air through the aerenchyma of the rice plants (Yagi 1997).

Iron is the most abundant member of the redox-sensitive elements in soil. Ferrous iron reacts with phosphate, carbonate, and sulfide to produce vivianite (see Sect. 5.3), siderite (see Sect. 5.4), and iron sulfide (see Sects. 5.5.1 and 5.5.2), respectively.

The occurrence of oxidizing conditions in the reduced paddy field soil depends on the management of irrigation water, as shown in Fig. 5.4. Oxidizing conditions after drainage can be detected most clearly from a lack of dipyridyl reaction and the presence of iron mottles (see Sect. 5.2).

Since the manganese content in soil is typically one-fifth or less of that of iron, it has limited effects on the morphological properties of Ap horizon soil, although faint and soft manganese concretions may be found in the paddy field subsoil. Nitrate is easily reduced to N_2 by denitrifiers, which affects the efficiency of N fertilizers, but does not affect the morphological properties of the Ap horizon soil.

5.1.3 Water Management and Characteristics of a Paddy Field Soil Profile

The area percentage of irrigated paddy fields is approximately 60% of the total paddy-field area in the world (Global Rice Science Partnership 2013). Figure 5.3 shows an example of seasonal changes in an irrigated paddy field in Miyagi Prefecture, Japan. In spring, farmers till the drained paddy field (Fig. 5.3a) and apply basal fertilizers (Fig. 5.3b). During this season, paddy fields are partially dried. After introducing water, puddling, and transplanting in late spring, rice plants are grown under submergence (Fig. 5.3c) until the maximum tillering stage in early summer. Following midseason drainage for 2 weeks, the paddy field is irrigated intermittently or submerged again (Fig. 5.3d) until the end of August. Then, irrigation is stopped and the paddy fields are dried for harvest (Fig. 5.3e).

The Eh value in the plow layer soil of paddy fields responds to the paddy field water management. Changes in Eh value were monitored using a platinum electrode connected to a reference electrode and a voltmeter. Figure 5.4 shows examples of Eh change with paddy field water management. Three water management schemes were carried out: midseason drainage and intermittent irrigation (Fig. 5.4a), midseason drainage and submergence (Fig. 5.4b), and continuous submergence (Fig. 5.4c).

At about 50 days after transplanting and submergence, the Eh values ranged around −0.1 to −0.2 V, suggesting that all three of the plow layer soils were nearly fully reduced due to continuous submergence. The Eh values of plots (a) and (b) increased with midseason drainage. Subsequently, the Eh value of plot (a) tended to be higher than those of the other plots due to intermittent irrigation, whereas the Eh value of plot (b) decreased to the same level as that of plot (c), which was continuously submerged. One hundred and ten days after transplanting, all the

Fig. 5.3 Seasonal changes in a paddy field. (**a**) After spring tillage and basal fertilizer application, (**b**) close-up view of applied fertilizers, (**c**) initial growth with submergence, (**d**) heading stage, (**e**) rice plant harvest time

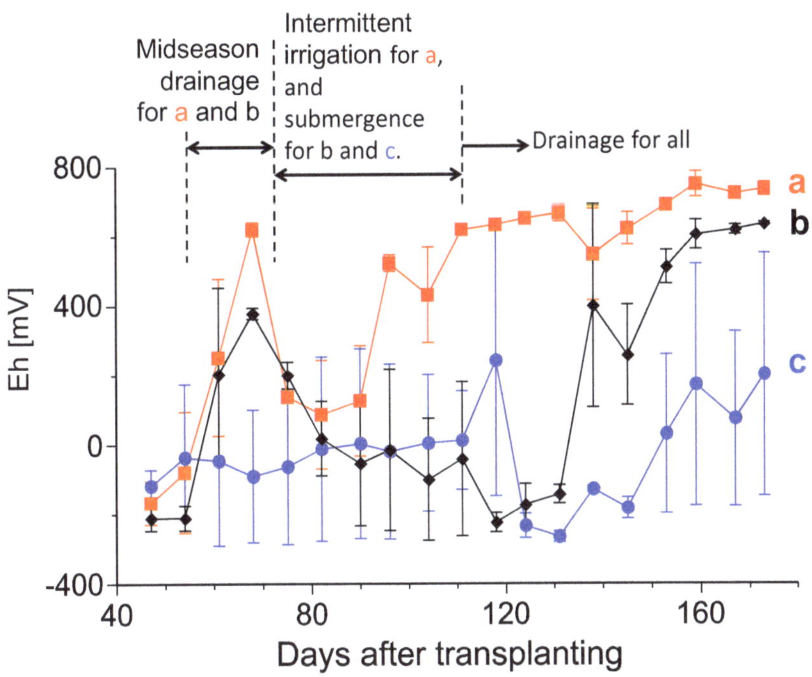

Fig. 5.4 Changes in soil redox potential values (Eh) with different water management schemes. (**a**) Midseason drainage and intermittent irrigation (MI), (**b**) midseason drainage and submergence (MS), (**c**) continuous submergence (CS). The error bars show the absolute deviation from the mean (n = 2)

plots were drained, as is usual in the water management practice of irrigated paddy fields. At the start of drainage, the Eh value of plot (a) was already higher than those of the other plots due to its treatment by midseason drainage followed by intermittent irrigation. The Eh value of plot (c) was the lowest, even after drainage started, due to continuous submergence, which may have delayed drainage compared to plot (b).

As a result of redox reactions caused by the water management schemes, the paddy field plow layer soils show periodic chemical changes every year. Directly under the plow layer soil, the plow pan, also common to paddy field soils, is formed. The plow pan is 5–10 cm thick, hard, and sometimes has weakly developed platy structure with hydrated iron oxide coating or mottles. In contrast, the properties of the lower horizon soils under the plow pan are mainly related to topography and ground water level, as well as related to redox reactions. Paddy fields can be divided into two types: well-drained and poorly drained (Kyuma et al. 1988; Wada and Neue 1988). Intermediate types also exist.

Well-drained paddy fields are distributed in relatively well-drained areas, such as uplands or natural levees in lowland areas. The lower horizon soils underlain by the plow pan are unsaturated with water and oxidative at least during the time when the paddy field is not submerged. As the plow layer soil is reduced under submergence with water permeability of approximately 2–3 cm d^{-1}, Mn^{2+} and Fe^{2+} are gradually transported to the subsoil, oxidized, and precipitated, forming Mn- and Fe-mottled and enriched layers. The Fe-enriched layer is more distinct and occurs shallower

Fig. 5.5 Poorly drained paddy field soil. (**a**) Two major types of iron mottles found in paddy field soils. (**b**) iron mottles resembling a coating of irregular or vesicular pore surfaces, (**c**) cylindrical iron mottles resembling root iron plaque

than the Mn-enriched layer (Wada and Neue 1988; Kyuma 2004), reflecting the difference in Eh^0(pH 7) values of the redox reactions related to these elements (Fig. 5.2).

Poorly drained paddy fields are distributed in marshlands or other areas with high ground water levels. The lower horizon soil directly under the plow layer is saturated with water and reduced, even after irrigation is stopped, due to high ground water level and low water permeability (Fig. 5.5a). The subsoil is bluish gray in color with faint enrichment of Mn and Fe (Kyuma 2004).

5.2 Hydrated Iron Oxide

Hydrated iron oxides are the most abundant redox-sensitive inorganic constituents in soil. Here pick up hydrated iron oxide of paddy field soils including iron mottles. The exemplified iron mottles correspond to redox concentrations described by Vepraskas (1992) and Hurt et al. (1996). Hydrated iron oxides here may range from noncrystalline hydrated iron oxide to ferrihydrite and poorly crystalline iron oxyhydroxides. Under oxidizing conditions, the Fe_o/dithionite-extractable iron (Fe_d) values of paddy field plow layer soils are higher than 0.3, suggesting that the major form of iron is poorly crystalline ferrihydrite due to yearly repetition of reduction and oxidation, i.e., dissolution and precipitation (Childs et al. 1991). Although lepidocrocite was identified in the iron mottles (Kojima 1971), lepidocrocite is partly soluble in acid oxalate solution (dark) (Schwertmann 1973; Fonseca and da Silva 1998). A significant form of ferrous iron is exchangeable or acetate-extractable Fe^{2+} (Kyuma 2004), and other forms of ferrous iron may include vivianite, siderite, and noncrystalline ferrous sulfide.

Figure 5.5a shows a profile of a poorly drained paddy field soil. The plow layer soil is gray in color, suggesting that the bulk soil is still under reducing conditions although irrigation had been stopped more than one month before, and the rice harvest was finished. As the texture of this soil is fine (Togami et al. 2017), with the major clay mineral being montmorillonite (Fig. 3.8), the soil is poorly drained. This soil profile contains two major types of iron mottles. The first type is brown-colored iron mottles, which resemble a coating of irregular or vesicular pore surfaces, as shown in Fig. 5.5b. Since the pore surface coating-like mottles are found in the plow layer soil, they were formed after puddling in the spring of that year. During the next rice cultivation, the mottles will be reduced and dissolved again. The second type is cylindrical iron mottles, which resemble root iron plaque. Since the layer rich in these mottles is deeper than 15 cm, the iron plaque-like mottles may have been formed by hygrophytes other than cultivated rice. A few small iron plaque-like mottles can also be found in the plow layer soil. Iron minerals included in the root iron plaque are suggested to be ferrihydrite, lepidocrocite, and others (Kahn et al. 2016). A possible reason for the remaining brown, iron plaque-like mottles in the reduced subsoil may be depletion of easily decomposable organic matter in this soil horizon. The properties of these mottles were further examined microscopically.

The cylindrical iron mottles, shown in Fig. 5.6, are well developed and can be separated from a clod. Figure 5.6a shows a cross-section of an air-dried clod, which contains root iron plaque-like mottles of different sizes and colors. The brown-colored larger mottles can be dug out from a field-moist clod, as shown in Fig. 5.6b, and were probably iron plaque formed around the roots of former vegetation. Figure 5.6c shows the gently washed iron plaque. It contains sand-size particles in the brown-colored area. The color distribution pattern is concentric; from the outside inward the color changes from yellowish brown to brown and gradually to light brown. Figure 5.6d shows a longitudinal section of the cylindrical iron mottle (Fig. 5.6c). The sand-sized particles and color distribution pattern observed in Fig. 5.6c can be ascertained in Fig. 5.6d. The distribution of sand-sized particles within the cylindrical iron mottle (Fig. 5.6c, d) suggests that oxygen transported through the plant roots diffused outside of the roots, oxidized ferrous iron around the roots, and precipitated hydrated iron oxides in the reduced soil environment. Over time, more ferrous iron probably diffused from the soil matrix and concentrated around the iron plaque, increasing its thickness. Similar and larger cylindrical root

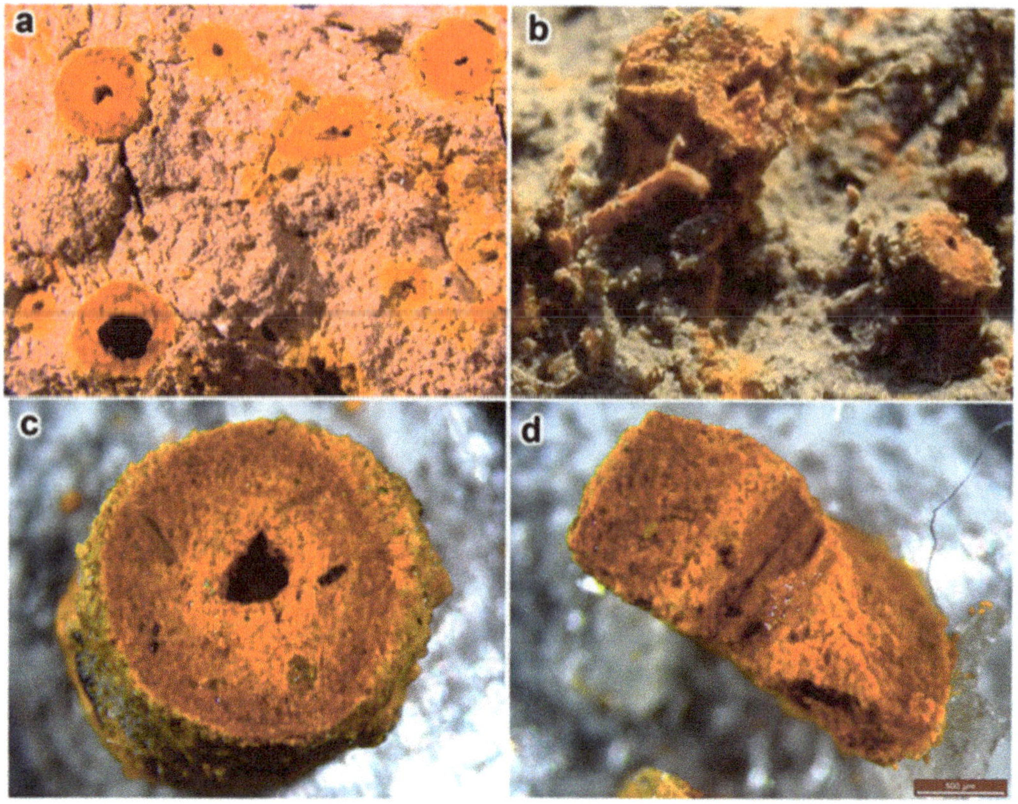

Fig. 5.6 Cylindrical iron mottles. (**a**) Iron mottles resembling root iron plaque, (**b**) digging the mottles out of the soil, (**c**) an iron mottle separated from the soil, (**d**) vertical section of the iron mottle in (c)

iron plaque has been reported from Pleistocene sediment in Aichi Prefecture, Japan (Yoshida and Matsuoka 2004).

A polished section was used to examine the elemental distribution in the cylindrical iron mottles. Figure 5.7a shows a cross-sectional scanning electron microscope (SEM) image of a cylindrical iron mottle, which appears as a bright and thick ring. The EDX spectra in Fig. 5.7b, c were obtained from the locations outlined in Fig. 5.7a by dashed squares (b) (cylindrical iron mottle) and (c) (bulk soil outside the iron mottle), respectively. The concentration of iron is much higher in the iron mottle than in the outer soil matrix. In contrast, the concentrations of Al and Si are similar, suggesting that the cylindrical iron mottle developed in the soil matrix directly outside the plant root. Element maps of Si (Fig. 5.7d) and Fe (Fig. 5.7e) support the interpretation of these EDX spectra. Although the Si concentration in the iron mottle appears to be slightly lower than that of the bulk soil, the distribution pattern of Si-rich particles in the iron mottle is the same as that of the bulk soil. The iron concentration in the iron mottle is significantly higher than that of the bulk soil, and the iron distribution pattern compares closely to the SEM image of the iron mottle (Fig. 5.7a). Potentially, the EDX spectrum for the iron mottle (Fig. 5.7b) has very small peaks of P and S that are lacking in the bulk soil EDX spectrum (Fig. 5.7c).

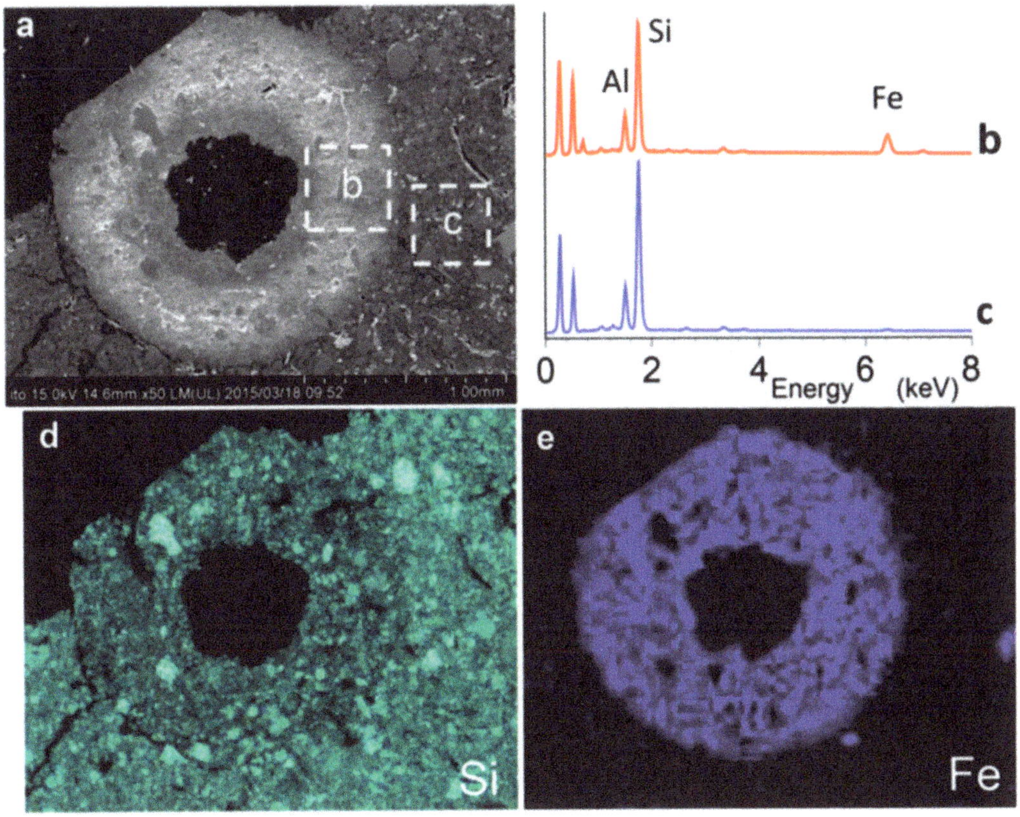

Fig. 5.7 SEM-EDX analyses of a cylindrical iron mottle using a polished section. (**a**) SEM image, (**b** and **c**) EDX spectra from the positions indicated by dashed squares (b) and (c), respectively, shown in (a), (**d** and **e**) Si and Fe element maps, respectively

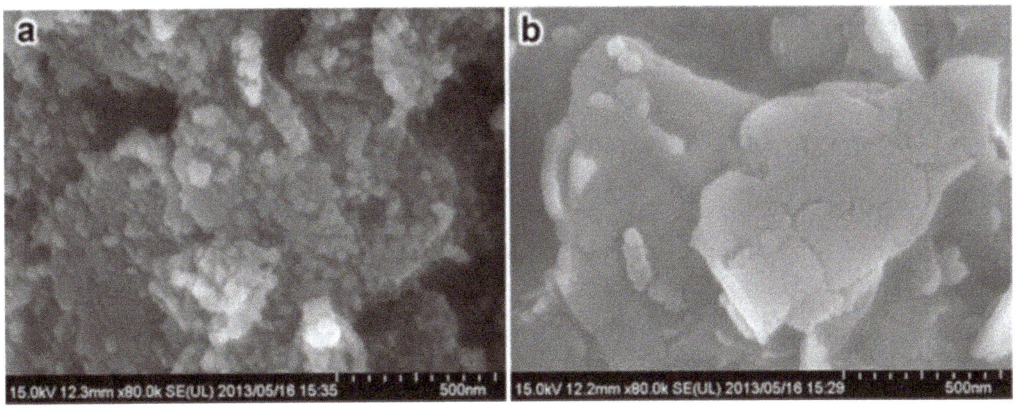

Fig. 5.8 Magnified SEM images. (**a**) cylindrical iron mottle, (**b**) bulk soil

As shown in Figs. 5.6 (color) and 5.7 (EDX spectra and Fe element map), differences in the distribution of iron between the iron mottle and bulk soil are evident. An additional difference between the mottle and bulk soil is observable in magnified SEM images (Fig. 5.8). Figure 5.8a shows that in the cylindrical iron mottle, aggregates of very fine spherical particles (poorly crystalline hydrated iron

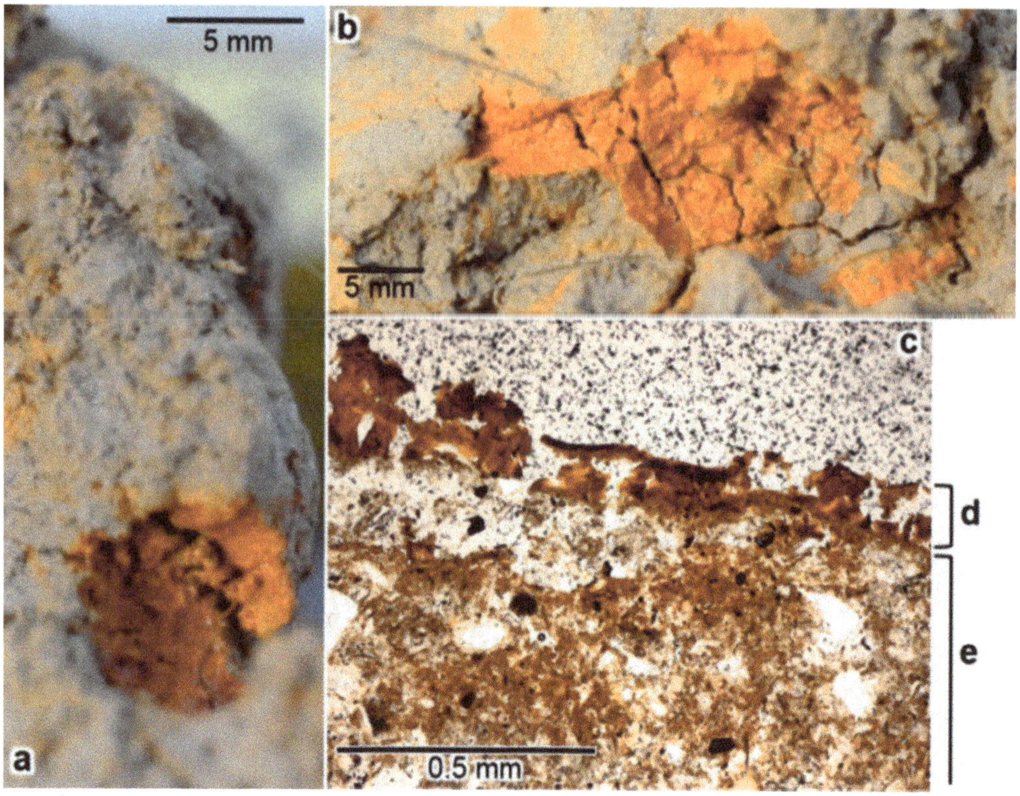

Fig. 5.9 Iron mottles coating the pore surfaces. (**a**) vesicular pore surface, (**b**) irregular pore surface, (**c**) thin section micrograph of the iron mottle (**d**) and bulk soil (**e**)

oxides) surround montmorillonite platy particles. In contrast, Fig. 5.8b shows that platy montmorillonite particles in the bulk soil lack these very fine spherical particles.

The next example of hydrated iron oxide taken from the plow layer soil (Fig. 5.5b) is iron mottles that resemble a coating of irregular or vesicular pore surfaces. Examples are shown in Fig. 5.9a (round vesicular shape) and (b) (irregular shape). These iron mottles, resembling a pore surface coating or lining, are more common inside the plow layer soil than on the soil surface where the soil is in direct contact with air. Considering the smoothly curved surfaces of these pores (Fig. 5.9a, b), they may have formed by gas (CO_2 or CH_4) production from microbial activity. Examining the thin section prepared to show a cross-section of the iron coating (Fig. 5.9c) under plane polarized light, the brown-colored part (Fig. 5.9d), which is the pore surface coating, contains no sand-sized particles. In contrast, the bulk soil (Fig. 5.9e) contains many transparent sand-sized particles. This observation suggests that the hydrated iron oxide coating the pore surfaces was formed outside the bulk soil. This occurrence of hydrated iron oxide is different from the cylindrical iron mottles that were formed in the bulk soil around plant roots (Figs. 5.6, 5.7, and 5.8), although the cylindrical mottles are thicker than the mottles coating the pore surfaces.

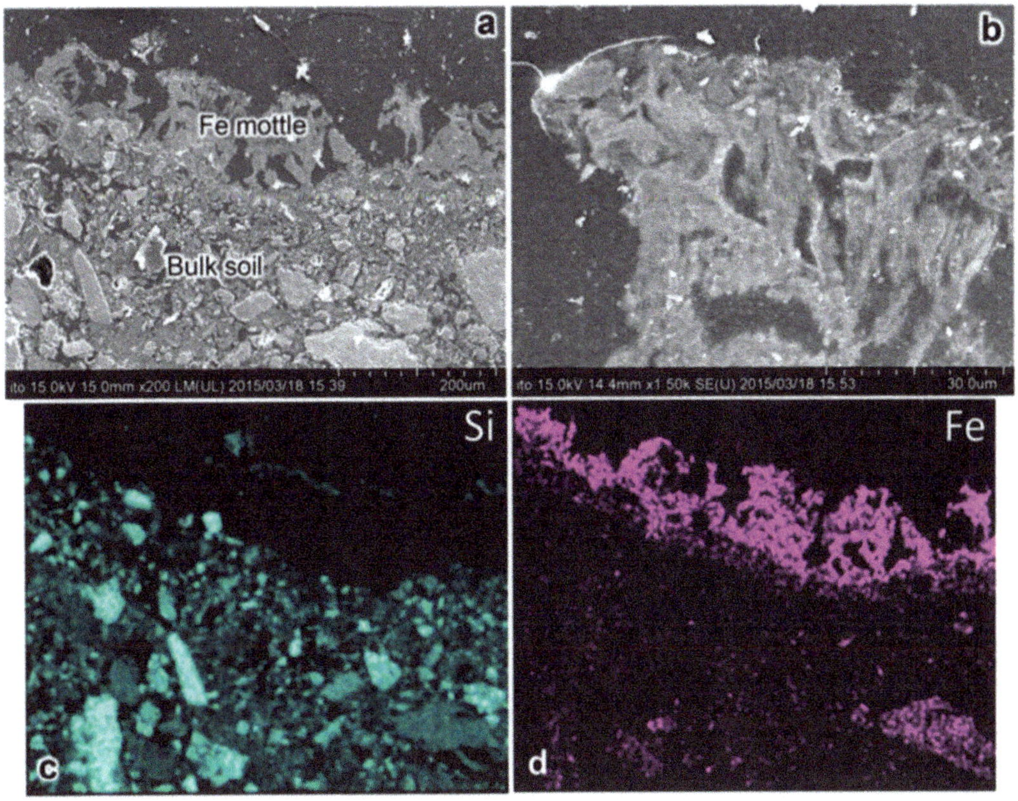

Fig. 5.10 Polished section of iron mottles coating the pore surfaces. (**a**) SEM image of polished section, (**b**) magnified SEM image of the iron mottle, (**c** and **d**) Si and Fe element maps, respectively

In order to further examine the elemental distribution in the mottles coating the pore surfaces, a polished section was prepared from a clod containing these mottles. Figure 5.10a shows a SEM image with similar magnification to the thin section (Fig. 5.9c). The thickness of the mottle is approximately 0.1 mm. Many sand-size particles are observed in the bulk soil area, whereas the mottle area has a streak-like structure. The streak-like structure of the mottle tends to be oriented from the soil surface to the pore space (Fig. 5.10b). This streak-like structure is somewhat different from the aggregates shown in Fig. 5.8a, although the magnification is not the same. The Si element map (Fig. 5.10c) shows a lack of Si-containing minerals in the iron mottle. The Fe element map (Fig. 5.10d) shows that the area of high Fe concentration compares very closely to the iron mottle in the SEM image (Fig. 5.10a).

The differences in structure between the cylindrical iron mottles and the iron mottles coating the pore surfaces may be due to the pore space at the redox interface where O_2 and Fe^{2+} meet. The former may have a barrier that prevents Fe^{2+} from diffusing smoothly inside the root. This barrier may be root sclerenchyma and/or exodermis. Consequently, oxygen diffuses into the bulk soil and hydrous iron oxide precipitates in the bulk soil. In contrast, the latter have space where hydrated iron oxide can precipitate outside the bulk soil. In the case of the iron mottles in the plow layer soil, phosphorus is also a significant constituent, as shown in Sect. 5.3.

5.3 Vivianite

As an iron phosphate mineral under reducing conditions, vivianite [$Fe_3(PO_4)_2 \cdot 6H_2O$] has been found in sediments or organic soils under reducing conditions (Rothe et al. 2016). Here we introduce the formation and dissolution of vivianite in the plow layer soil of paddy fields. Vivianite is formed and dissolved according to the redox conditions of the soil. Vivianite is formed significantly in ordinarily P-enriched paddy field soils, whereas it is not formed in Andisol paddy fields.

5.3.1 Detection of Vivianite in Paddy Field Soil

Vivianite has previously been reported in the deep horizon, 1 m from the surface, of paddy field soil with high ground water level (Ito 1975). This site of vivianite presence is too deep to be related to rice cultivation, i.e., P fertilizer application in the paddy field. Subsequently, Wada et al. (1977) reported a vivianite-like material on rice roots using an optical microscope. Their finding was directly related to rice plants. Their observation was later confirmed using X-ray microdiffraction and SEM-EDX analyses on rice roots grown in pots (Nanzyo et al. 2010) and in ordinary paddy fields (Nanzyo et al. 2013). Figure 5.11a shows vivianite crystal aggregates formed on rice roots. The color is blue–green, which is due to partial oxidation of iron after exposure to air. Regarding the color change of vivianite after exposure to air, see Fig. 5.20. The powder XRD pattern obtained from the vivianite aggregates (Fig. 5.11b) by the microdiffraction method is significantly different compared to that obtained from a rice root with almost no vivianite crystals. The XRD pattern shown in Fig. 5.11b is identical to the reference XRD pattern for vivianite reported by Lehr et al. (1967) (Fig. 5.11c). Hence, the XRD analyses show that vivianite is dominant in the crystal aggregates (Fig. 5.11). Important conditions required to detect vivianite on rice roots are (i) lowland paddy field soils (not Andisols), (ii) duration of approximately one-and-a-half months after transplanting a rice seedling under continuous submergence, (iii) plant-available P level of plow layer soil higher than approximately 0.1 g P_2O_5 kg^{-1} as determined by the Truog method, (iv) washing of rice roots to remove soil soon after sampling, (v) air-drying of washed roots, and (vi) magnifying glass with magnification factor of greater than 30 times.

The elemental composition of a vivianite crystal aggregate was examined (Fig. 5.12). From the crystal aggregate (Fig. 5.12a), dashed square b was selected, and the corresponding EDX spectrum (Fig. 5.12b) showed that the crystal aggregate

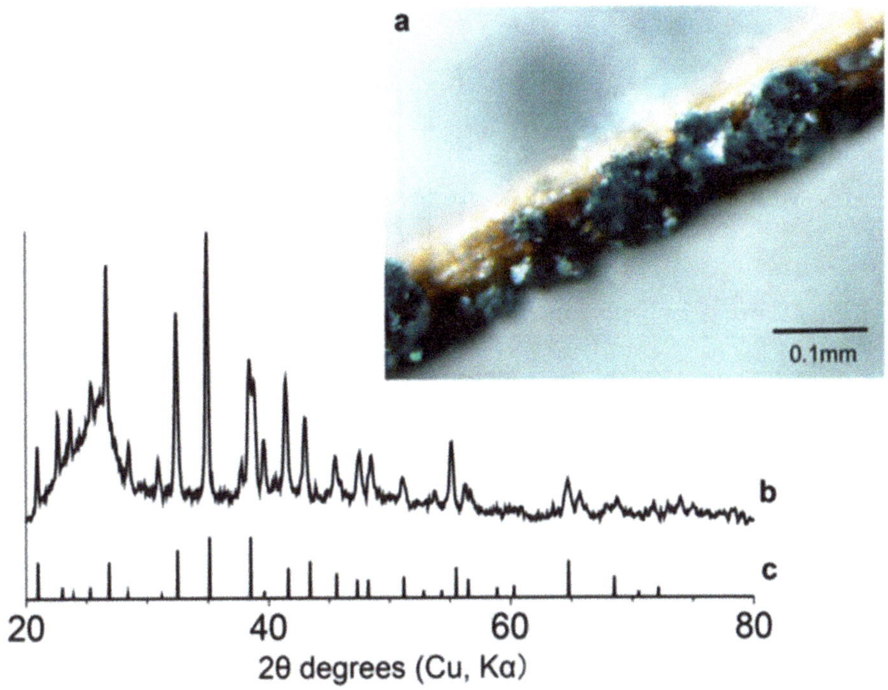

Fig. 5.11 Detection of vivianite by X-ray microdiffraction. (**a**) optical micrograph of vivianite formed on rice roots, (**b**) XRD pattern of vivianite crystal aggregates, (**c**) reference XRD pattern of vivianite (Lehr et al. 1967)

dominantly consists of Fe and P. Element maps of P (Fig. 5.12c) and Fe (Fig. 5.12d) compare closely to the SEM image (Fig. 5.12a), indicating that the crystal aggregate surrounding the root in Fig. 5.12a is dominantly composed of vivianite.

Although the surfaces of rice roots in the reduced paddy field soil are covered with oxidized iron plaque, some of this plaque may be reduced with aging. During plaque formation on rice roots, P released by reduction of hydrated iron oxides in the soil can be sorbed and accumulated on the iron plaque, as described later in Sect. 5.3.3. With an increase in reducing conditions to the iron plaque, vivianite is formed. Since new rice roots develop in succession with an increase in tillering, and the redox conditions in the soil around roots may likely vary, for example, with depletion of easily decomposable organic matter, some iron plaque may remain until after the rice harvest.

As a result, considering that rice roots comprise a mixture of young and old specimens, the ratio of P in the vivianite form is only half of the P contained in rice roots (Nanzyo et al. 2013). The other half may be iron phosphate material that exists at the redox interface cells shown in Figs. 5.17 and 5.19, possibly the root scleren-chymatous layer and/or exodermis.

Under reducing conditions in soil, the content of ferrous salts other than vivianite may be considerable. The other ferrous salts are $Fe(OH)_2$, siderite, and noncrystalline ferrous sulfide. Figure 5.13 compares the stability of these four ferrous salts in an approximately neutral pH range under hypothetical conditions (Nanzyo et al. 2010).

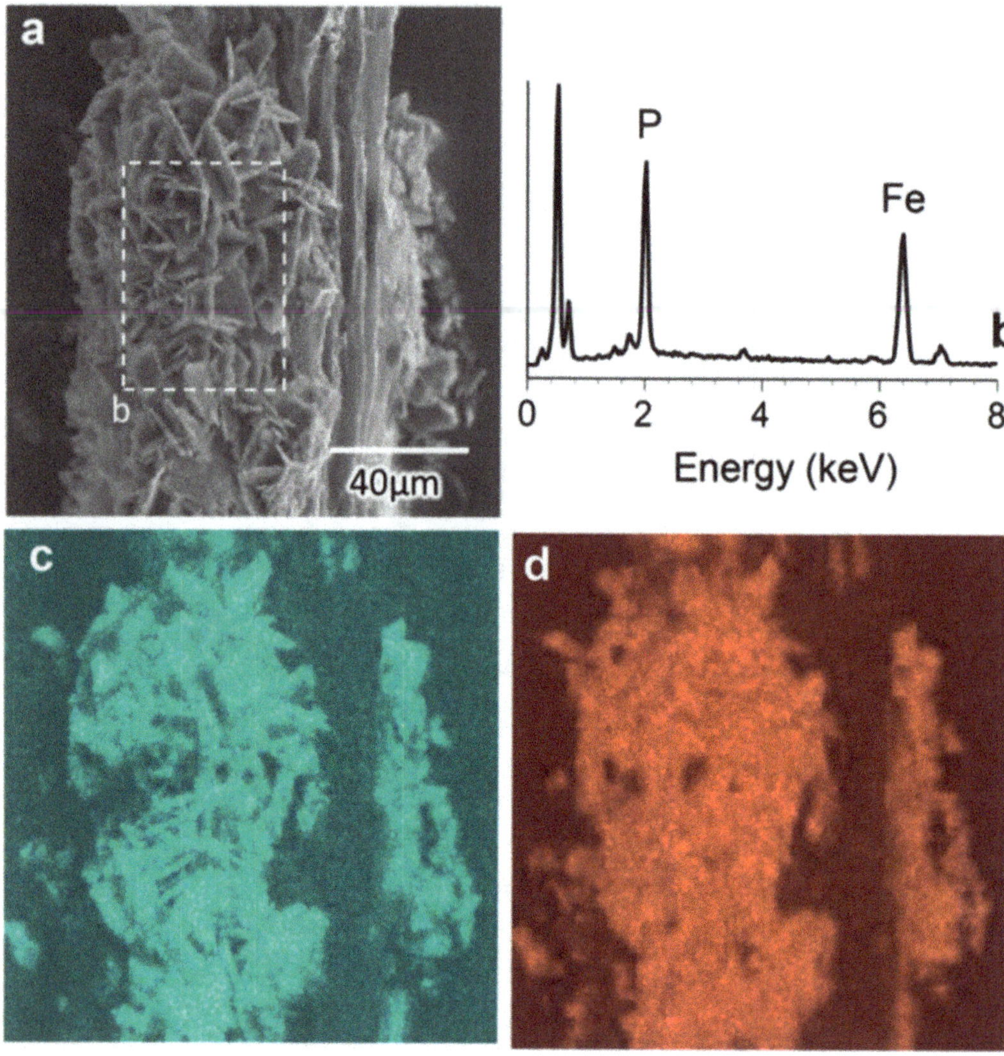

Fig. 5.12 Elemental composition of vivianite on a rice root. (**a**) SEM image of vivianite crystal aggregates formed on a rice root, (**b**) EDX spectrum for dashed square (b) shown in (a), (**c**, and **d**) P and Fe element maps of the crystal aggregates, respectively

In this diagram, a solid phase with a higher log $\{A_i\}/\{Fe^{2+}\}$ value is considered to be more stable than those with lower values, where $\{A_i\}$ shows the formal activities of the solid phase in the same way as the activities of the solutes. Uncorrected equilibrium constants (ionic strength = 0) were used for simplicity. The concentrations of total dissolved phosphorus (P_T), sulfur (S_T), and inorganic carbon (C_T) were tentatively assumed to be 10^{-6}, 10^{-6}, and 10^{-3} mol L^{-1}, respectively. These values may vary widely with soil conditions and biological activity (Kirk 2004; Kyuma 2004), which may affect the results. Furthermore, the reported solubility product value for vivianite can vary by several orders of magnitude (10^{-32}–10^{-36}) (Sadiq and Lindsay 1979; Stumm and Morgan 1996). This discussion about the relationship of these ferrous salts is made under these assumptions. The stability of FeS (amorphous) is the highest of the four salts. However, since the labile S content in the studied soil was

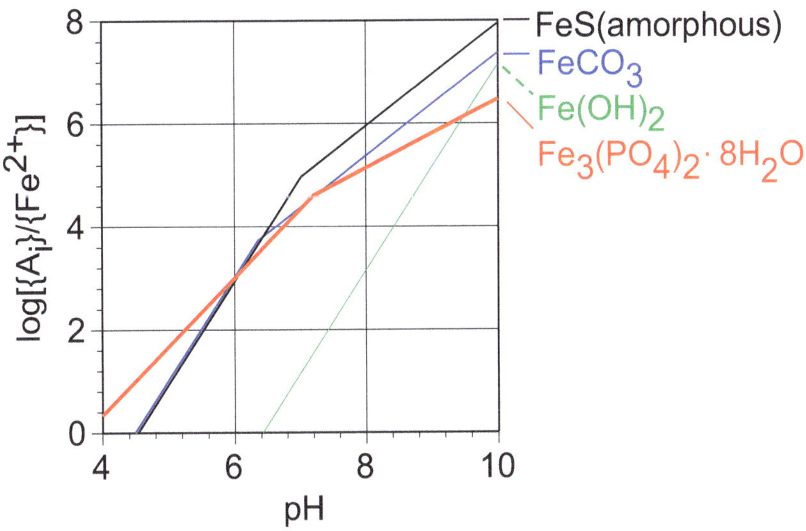

Fig. 5.13 Activity ratio diagram for $P_T = S_T = 10^{-6}$ mol L^{-1} and $C_T = 10^{-3}$ mol L^{-1} in a Fe–P–S–C system. $\{A_i\}$ shows the activities of the solid phase. Amorphous FeS and vivianite have been added to the activity ratio diagram for $Fe(OH)_2$ and $FeCO_3$ described by Stumm and Morgan (1996)

much less than Fe_o, enough ferrous iron was still available for the formation of other solid phases, such as $FeCO_3$ and vivianite, in the neutral soil pH range under reducing conditions. The activity ratios of the latter two minerals are shown to be relatively similar in Fig. 5.13. Both of these minerals may form if enough ferrous iron is still available for the mineral with the lower stability. Thus, our present observation of solid phases in the reduced soil is compatible with reported thermodynamic data and stability relationships.

5.3.2 *Effect of Water Management on Vivianite in Paddy Field Soil*

Indices of biologically available P, such as Truog P or Bray No. 2 P, and dissolved P levels in soil water have been reported to increase under reducing conditions compared to those under oxidizing conditions (Shiga and Yamaguchi 1976; Kyuma 2004). Previous research has suggested that vivianite may be responsible for these observations. Heiberg et al. (2012) also considered vivianite to explain P behavior in soils under reduced conditions. Considering thermodynamic stability, vivianite is thought to form under reducing conditions and to dissolve after soil oxidation (Lindsay 1979); therefore, the vivianite content is most likely affected by the changing redox conditions in paddy field soils between submergence and drainage. Thus, the vivianite content of rice roots was examined under the three different water management schemes shown in Fig. 5.4.

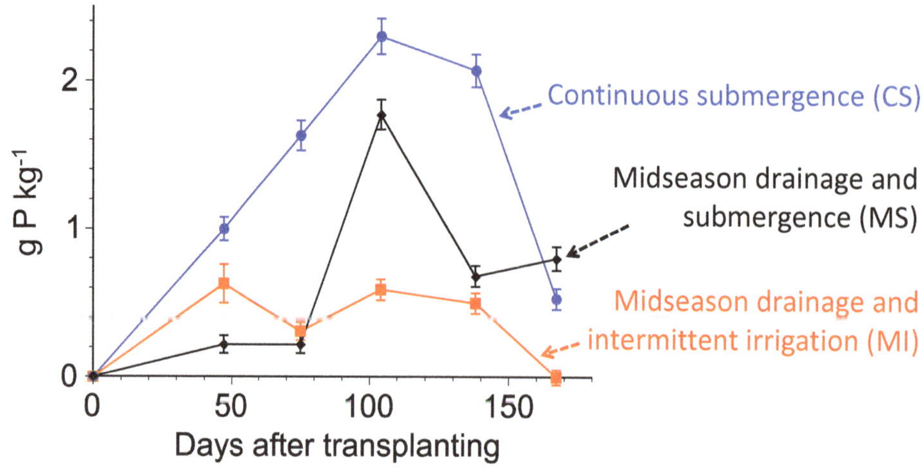

Fig. 5.14 Changes in the vivianite content of rice roots with three different water management schemes. The vivianite content of rice roots was estimated based on its property that the solubility of vivianite in mixed acid solution (0.1 mol L^{-1} HCl and 1 mol L^{-1} acetic acid) greatly decreases after heating at 105 °C (Nanzyo et al. 2013). The error bars show the absolute deviation from the mean (n = 2)

Different water management schemes affected the vivianite content on rice roots (Fig. 5.14). The highest vivianite content was present in the continuous submergence (CS) plot. Following drainage, the vivianite content decreased with oxidation of the plow layer soil. The slow decrease in the vivianite content of the CS plot was possibly related to slow drainage and slow oxidation due to the lack of midseason drainage. In the midseason drainage and submergence (MS) plot, the vivianite content increased to a level almost similar to that of the CS plot at 104 days after transplanting (Fig. 5.14). Of the three plots, the vivianite content was lowest in the midseason drainage and intermittent irrigation (MI) plot. The Eh value in the MI plot tended to be higher following midseason drainage compared to the other plots (Fig. 5.4). The observed changes in the vivianite content on rice roots with water management scheme (Fig. 5.14) followed the changes in the Eh value (Fig. 5.4). These observations indicate that vivianite formation is strongly affected by water management and the redox conditions in the plow layer soil of paddy fields. According to Lindsay (1979), vivianite dissolution is facilitated by a decrease in ferrous iron concentration, and the generation of P sorption sites by the precipitation of hydrated iron oxides.

Considering Lindsay's interpretation about the behavior of vivianite after the cease of irrigation and subsequent oxidation of the plow layer soil of a paddy field, the P content of iron mottles in the plow layer soil was examined. Figure 5.15a shows part of a polished section of a clod with iron mottles coating the pore surfaces. The form of the iron mottle is similar to that shown in Figs. 5.9 and 5.10, the color is brown, and the thickness is 0.05–0.15 mm. Figure 5.15b–d shows the element maps of P, Fe, and Si obtained from the dashed square area defined in Fig. 5.15e. The iron mottle layer contains P and Fe but no Si, suggesting that P was sorbed by hydrated iron oxide that precipitated at the pore surface space, as observed in Figs. 5.9 and

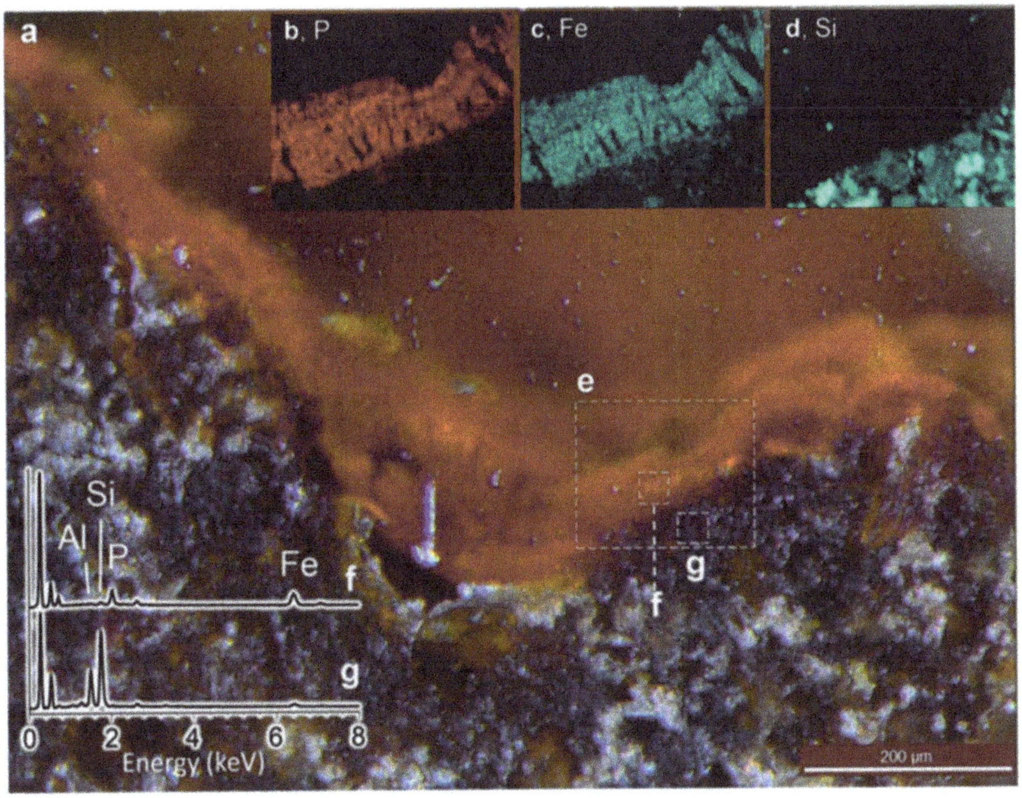

Fig. 5.15 Distribution of P in the iron mottles and bulk soil. (**a**) polished section of soil clod with iron mottles coating the irregular pore surfaces, (**b**, **c**, and **d**) P, Fe, and Si element maps for the dashed square area defined in (**e**), (**f**,and **g**) EDX spectra obtained from the dashed square areas (f) and g, respectively

5.10. There is also a streak-like structure in the iron mottle. The EDX spectra obtained for the dashed square areas (f) (iron mottle) and (g) (bulk soil) in Fig. 5.15 show that a significant content of P was detected in the iron mottle, whereas the P content in the bulk soil was very low. Hence, the iron mottle coating on the pore surfaces forms at least one of the P sink sites after dissolution of vivianite associated with soil oxidation.

5.3.3 P Accumulation at Redox Interfaces of Rice Roots

Cylindrical iron mottles or iron plaque developed on the roots of a hygrophytes introduced in Sect. 5.2 contained only a very small amount of P (Fig. 5.7b). Since the soil horizon was deeper than the plow layer soil, the effect of P fertilizer application was small. However, in the plow layer soil, the plant-available P level is high due to P fertilizer application. The rice roots are macroscopically brown in color (Fig. 5.16a), and iron plaque partly remains on the roots 1 month after harvest. From the washed and dried rice roots (Fig. 5.16a), a bundle was cut 5–7 cm from the root base, and a polished section was prepared to examine a cross-section of the iron

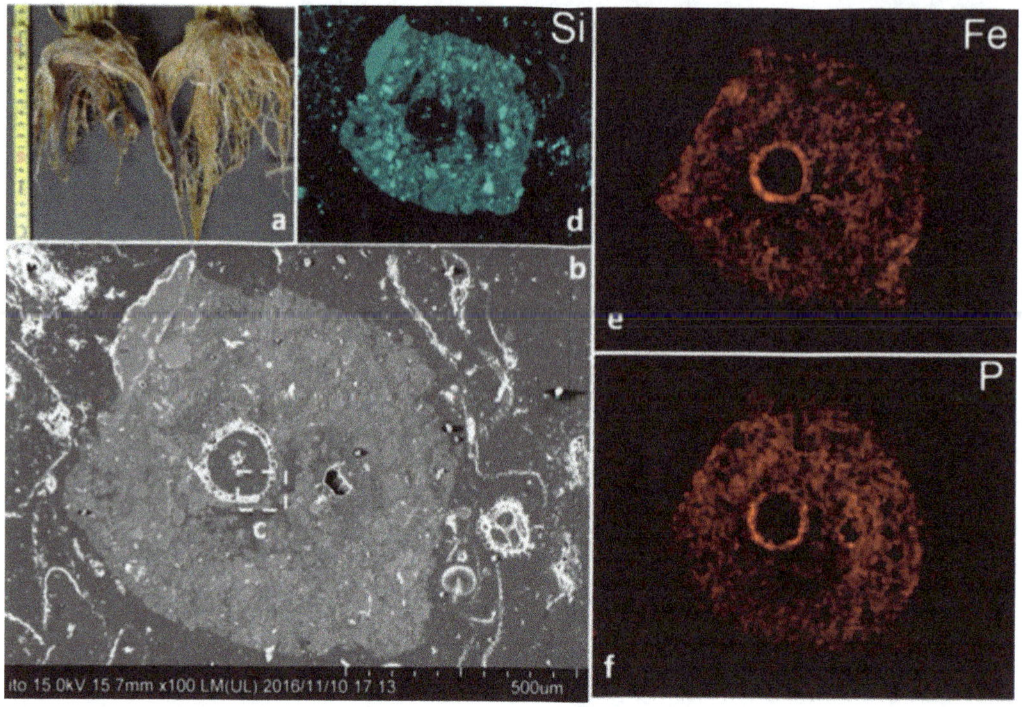

Fig. 5.16 Distribution of P in root iron plaque from the plow layer soil of a paddy field. (**a**) Rice roots washed out of the soil after harvest, (**b**) SEM image of polished cross-section of a rice root containing iron plaque, (**c**) dashed square area for Fig. 5.17, (**d**, **e**, and **f**) element maps for Si, Fe, and P, respectively

plaque. Figure 5.16b shows a SEM image of the rice root iron plaque. At the center, there is a thin rice root, of which only the exodermis-like layer and stele are likely to remain. Element maps of Si (Fig. 5.16d) and Fe (Fig. 5.16e) show that the iron plaque developed in the bulk soil around the rice root. These characteristics are similar to those shown in Fig. 5.7d, e. However, a difference observed in the Fe element map (Fig. 5.16e) is that the iron concentration in the exodermis-like layer is higher than that in the iron plaque developed outside the exodermis-like layer. Examining the P element map (Fig. 5.16f) shows that a P distribution pattern closely related to that for Fe was obtained. Hence, the iron plaque is the site of P accumulation in the plow layer soil, and these Fe and P accumulations may be converted to vivianite if the iron plaque is reduced. Remarkably, both Fe and P are highly accumulated in the exodermis-like layer in this example. Therefore, an area of the exodermis-like layer (dashed square in Fig. 5.16c) was magnified in Fig. 5.17a.

The EDX spectrum of the material inside the exodermis-like cells shows that it is similar to iron phosphate (Fig. 5.17b, Nanzyo et al. 2004). Since the material in each cell is localized toward the inner part of the root, the oxidizing power originated from within the root, possibly through the intercellular space. Based on the cell diameter, 10–17 μm, these cells are exodermis rather than sclerenchyma (Kondo et al. 2000). The intracellular material also contains a small amount of Ca, but almost no Si. Thus, the inner parts of the exodermis-like cells may be under a different environment from

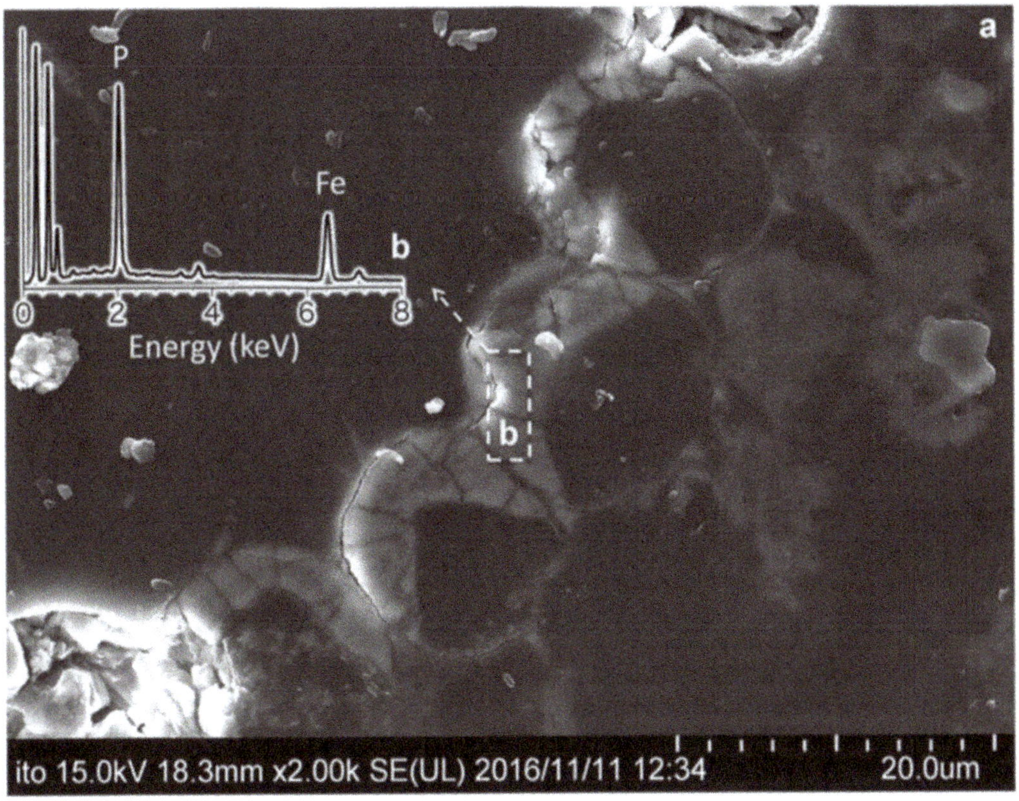

Fig. 5.17 Iron phosphate material in the exodermis-like cells. (**a**) magnified SEM image of the dashed rectangular area in Fig. 5.16c, (**b**) EDX spectrum obtained from the dashed rectangular area (**b**), shown in (**a**)

the outer parts, which is the site of iron plaque formation. The iron phosphate-like material is also different from vivianite because it does not show a blue color under the optical microscope. It was likely formed at the redox interface between the oxidative intercellular space inside the root and the reductive bulk soil outside the root. This iron phosphate-like material may be one of the reasons why the vivianite content of rice roots is as high as 50% even at the end of the paddy field submergence period (Nanzyo et al. 2013).

Figures 5.16 and 5.17 exhibit the thick iron plaque. However, when the rice roots are washed out of the soil, the number of brown roots without thick iron plaque is much greater than those with thick iron plaque. Figure 5.18b shows an optical microscope photograph of the surface of a thick rice root, which displays a brown rod-like color pattern. Although this pattern is not observable with SEM, it is observable with EDX, as shown in the Fe element map (Fig. 5.18c) for the dashed square area outlined in Fig. 5.18d. The characteristic X-ray used for the EDX analyses provides information about deeper sites than the secondary electron beam used for SEM. These observations suggest that the brown rods are located under the exodermis cell walls. A similar pattern can be obtained for P, but it is less clear than that for Fe, suggesting that the Fe concentration in the brown rods is higher than that

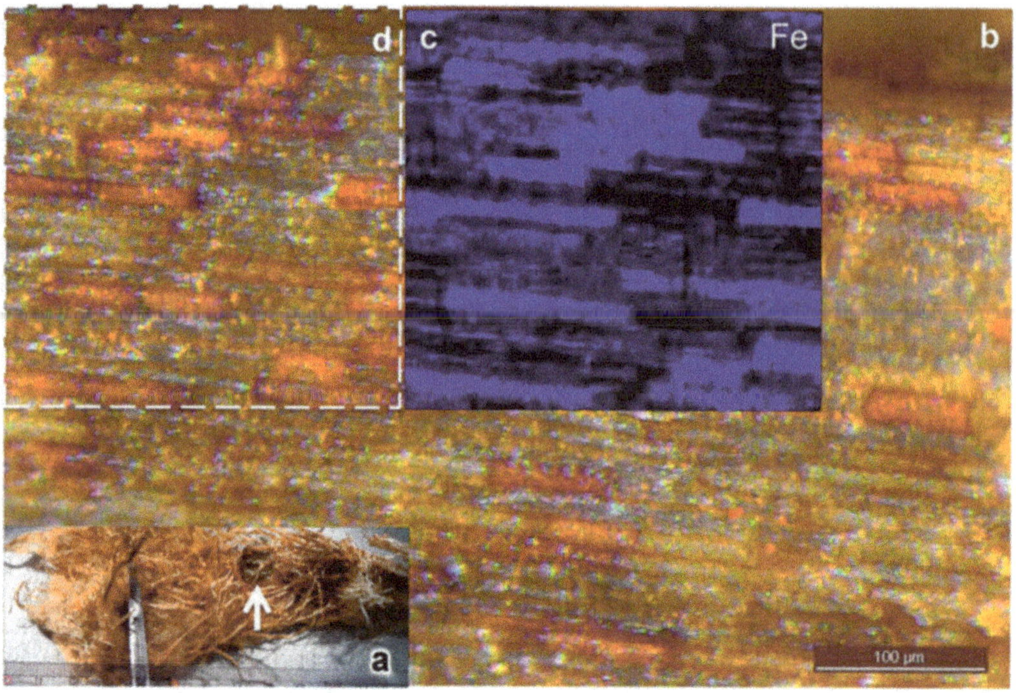

Fig. 5.18 Analysis of a brown thick rice root after rice harvest. (**a**) Air-dried rice roots after harvest with an arrow showing the position of the sampled root piece, (**b**) optical micrograph of the rice root piece, (**c**) Fe element map for the dashed square area in (**d**)

of P. In order to examine these brown rods, the roots were treated with hot H_2O_2 to digest the cell walls.

Under the optical microscope, the digested roots showed brown rods, and light brown-colored or nearly transparent particles with irregular shapes (Fig. 5.19a). After selecting a dashed square area (Fig. 5.19b), a SEM image (Fig. 5.19c) as well as element maps of Fe (Fig. 5.19d), Si (Fig. 5.19e), and P (Fig. 5.19f) were obtained. Integrating these results showed that the brown rods and particles are rich in Fe and P, suggesting that they are composed of hydrated iron oxide and sorbed phosphate. Hence, it is probable that the brown rods observed on the root surface (Fig. 5.18b) are the brown rods observed in Fig. 5.19a. These brown rods were formed at the interface between oxidizing conditions inside the roots and reducing conditions outside the roots. The diameter (15 µm) of the large brown rods suggests that they were formed inside the exodermis cells (Kondo et al. 2000). The longevity of P-rich hydrated iron oxide inside the sclerenchymatous cells and/or exodermis cells, and iron plaque, may depend on the intensity of aeration through the intercellular space, and the reducing conditions in the soil. Nearly transparent particles observed in Fig. 5.19a, b appear to be phytoliths.

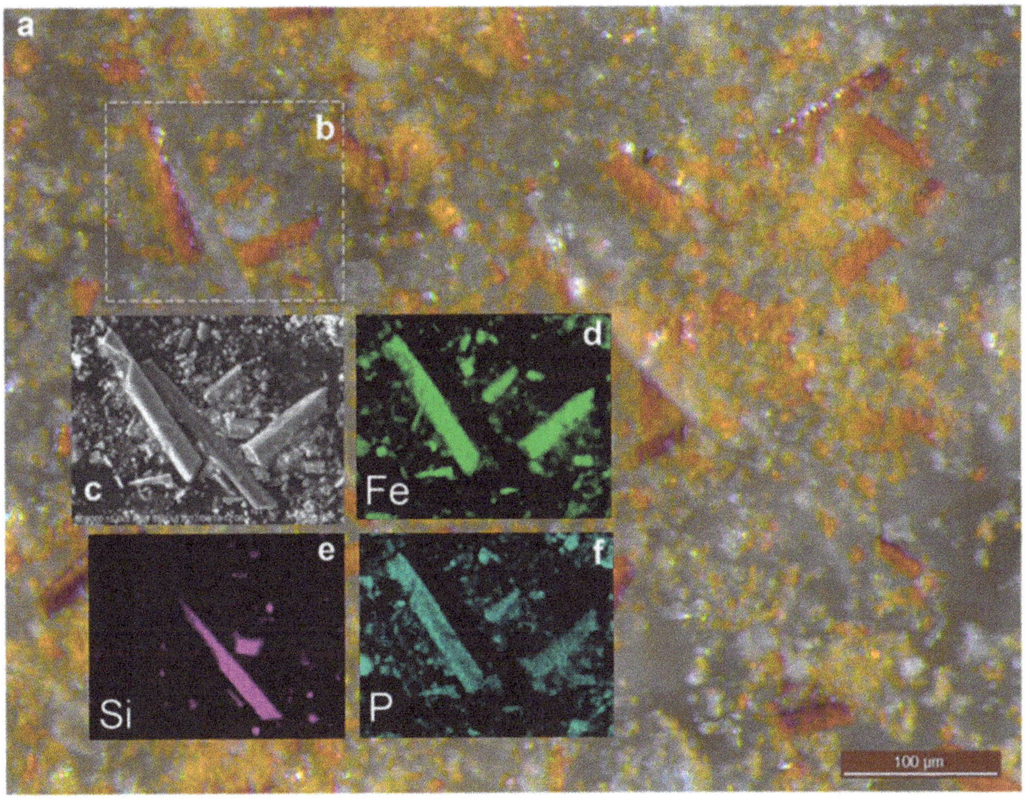

Fig. 5.19 Analyses of brown rice roots after rice harvest. (**a**) Optical micrograph of H_2O_2 digestion residue of rice roots, (**b**) dashed square area chosen for SEM-EDX analyses, (**c**) SEM image of the area outlined in (**b**), (**d**, **e**, and **f**) element maps for Fe, Si, and P, respectively

5.3.4 Vivianite Formation in Bulk Soil

Figures 5.11 and 5.12 showed vivianite formation on rice roots. An advantage of using rice roots for vivianite detection is the simplicity of separating vivianite from soil. In addition, iron plaque around rice roots was the site of P accumulation under reducing soil conditions, which is preferable for vivianite formation, as discussed above. Incidentally, the occurrence of vivianite crystals on the rice roots suggests that parts of them are growing from exodermis-like cells toward the bulk soil. It appears that vivianite can be formed in bulk soil (Zelibor et al. 1988). In fact, vivianite crystals form aggregates larger than 0.05 mm in diameter. By using soil with a particle size of less than 0.038 mm, vivianite crystal aggregates can be separated from the fine soil fraction using a 0.053 mm sieve.

Figure 5.20a shows vivianite crystal aggregates separated from the plow layer soil of a paddy field under submergence after incubation for 30 days at 30 °C. Vivianite crystals are colorless immediately after separation from the incubated soil. Subsequently, the color of the vivianite crystals slowly turns blue (Fig. 5.20b) by partial oxidation of Fe^{2+} in air (Garnd and Lavkulich 1980). Since other soil minerals are included in the crystal aggregates, the vivianite content of these aggregates is approximately 40%.

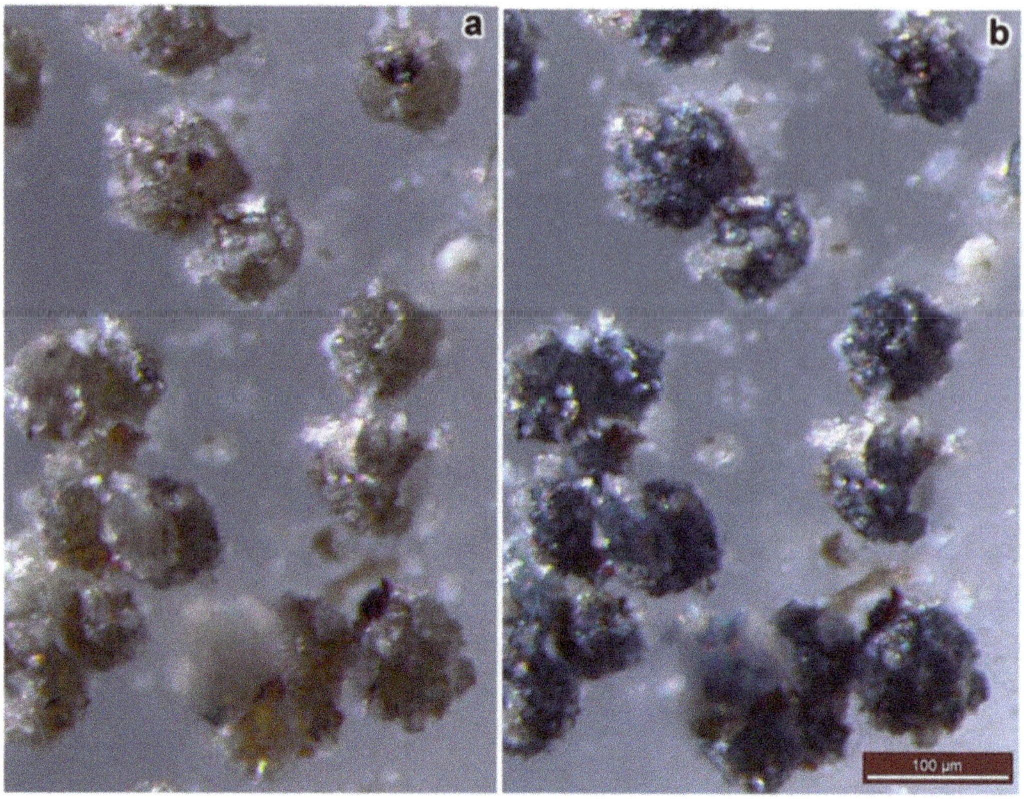

Fig. 5.20 Optical micrographs of vivianite crystal aggregates. (**a**) Immediately after separation from soil incubated for 30 days at 30 °C, (**b**) after allowing to stand for 8 h at approximately 25 °C

Vivianite forms in bulk soil under reducing conditions. Vivianite may affect the P dynamics in the plow layer soil of paddy fields (Walpersdorf et al. 2013). In a lowland paddy field soil, the Fe_o content is typically much higher than Al_o. Under oxidizing conditions, active Fe materials in the plow layer soil play a major role in sorbing P from fertilizers. The P concentration in the soil solution increases with soil incubation under submergence (Shiga 1973). In that previous study, although vivianite formation was not confirmed, it is probable that the increase in P concentration was related to vivianite formation. However, the formation of vivianite contributes to retain P in the plow layer soil of paddy fields after releasing P from active Fe materials.

It is believed to be effective to study the factors affecting the formation and dissolution of vivianite to control P dynamics, and to increase P efficiency, in the plow layer soil of paddy fields. Although to date it has been difficult to artificially extract P from paddy fields, vivianite can at least theoretically be separated from soil. Magnets can also be used to collect vivianite aggregates from incubated paddy field soil.

Fig. 5.21 Schematic P cycle in paddy fields of lowland

5.3.5 P Cycle in Irrigated Lowland Paddy Field Soil

The above-mentioned redox reactions related to P are outlined as the P cycle of rice production in Fig. 5.21. Starting with P fertilizer application to the plow layer soil (Fig. 5.21, left), the applied P is sorbed by active Fe minerals. Under submergence, P is partly absorbed by rice plants, and remaining reductant-soluble P is mainly converted to vivianite (Fig. 5.21, lower center), which is rice plant-available. Part of the P can be stored as P-rich hydrated iron oxides in sclerenchymatous and/or exodermis cells, and iron plaque. After irrigation is ceased, vivianite is dissolved and P is sorbed by hydrated iron oxides to give P-rich hydrated iron hydroxide and iron mottles (Fig. 5.21, lower right). Approximately three-quarters of the P absorbed by rice plants is transferred to rice grain, and used as food for humans or feedstuff for animal production. The P_2O_5 concentration in chicken manure and swine manure is as high as 5–6% on a dry-weight basis (Komiyama et al. 2013). The major constituent form of P is $MgNH_4PO_4{\cdot}6H_2O$ (see Sect. 6.3), which is also available for rice plants, although the application of compost to paddy fields is uncommon in Japan at present. Rice straws are, in many cases, cut, sprayed on the soil surface, and then plowed into the soil. In this way, the remaining P in the rice straws and roots is returned to the plow layer soil.

5.4 Siderite

Siderite [$FeCO_3$] is sometimes found in the subsoil of poorly drained paddy fields as white soft or firm nodules. Relatively large siderite nodules were found in a paddy rice field in Miyagi Prefecture, Japan (Fig. 5.22a). Since the paddy field was used for soybean cultivation at the time of sampling, the plow layer soil was brown in color. At depths of 15 cm or deeper, the soil color was bluish gray, indicating reduced conditions. Siderite was found at a depth of 40–50 cm below the surface (indicated

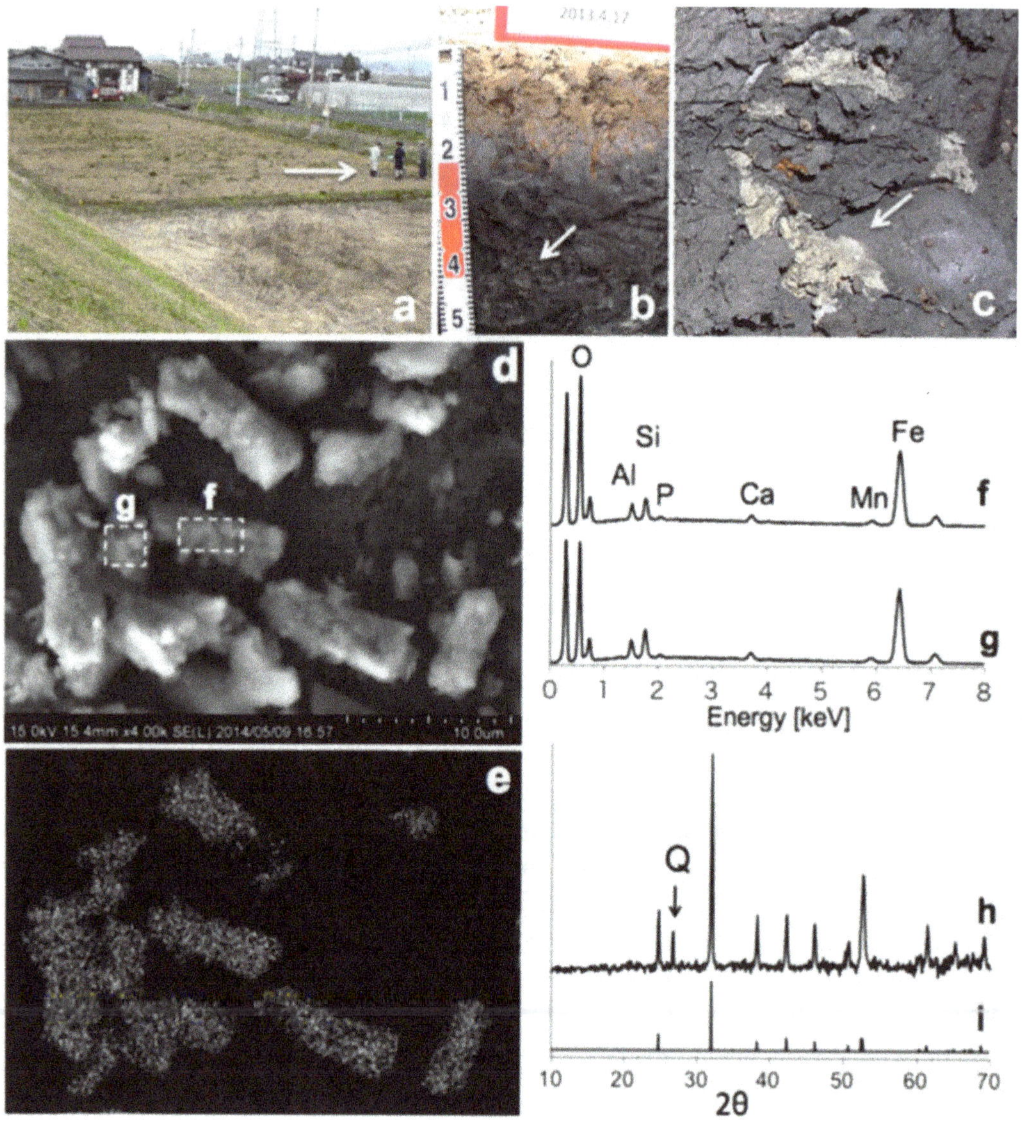

Fig. 5.22 Siderite in the subsoil of a poorly drained paddy field. (**a**) Sampling site (arrow shows the pedon site), (**b**) soil profile (arrow shows a siderite-rich horizon), (**c**) close-up of the siderite-rich horizon (arrow shows an example of a siderite-rich patch), (**d**) SEM image, (**e**) element map of Fe, (**f** and **g**) EDX spectra for selected areas (**f**) and (**g**) shown in (**d**), (**h**) XRD pattern of siderite from the paddy field subsoil where Q denotes a diffraction peak from quartz, (**i**) reference XRD data from Brindley and Brown (1980)

by the arrow in Fig. 5.22b). Magnified, the siderite-rich part was identified as firm gray-colored concretions, as shown in Fig. 5.22c. The siderite concretion was composed of an aggregate of small rod-like particles (Fig. 5.22d), as supported by the Fe element map (Fig. 5.22e). EDX spectra (Fig. 5.22f, g) were obtained from the dashed square areas identified in Fig. 5.22d. They show that Fe is the major element other than C and O, which are also elemental components of siderite.

The powder XRD pattern of the siderite concretion (Fig. 5.22h) is similar to that of reference siderite (Fig. 5.22i), although a small amount of quartz was also detected. The SEM observation also showed platy clay particles around the siderite (Togami et al. 2017), suggesting that the siderite nodules grew in the bulk soil.

Siderite in forms other than that shown in Fig. 5.22c, d were found in the same soil horizon, where it appears that siderite is filling the pores of root iron plaque. Figure 5.23a shows a SEM image of one example, and Fig. 5.23b, c shows the Si and Fe element maps, respectively. The EDX spectra (Fig. 5.23d, e) were obtained from

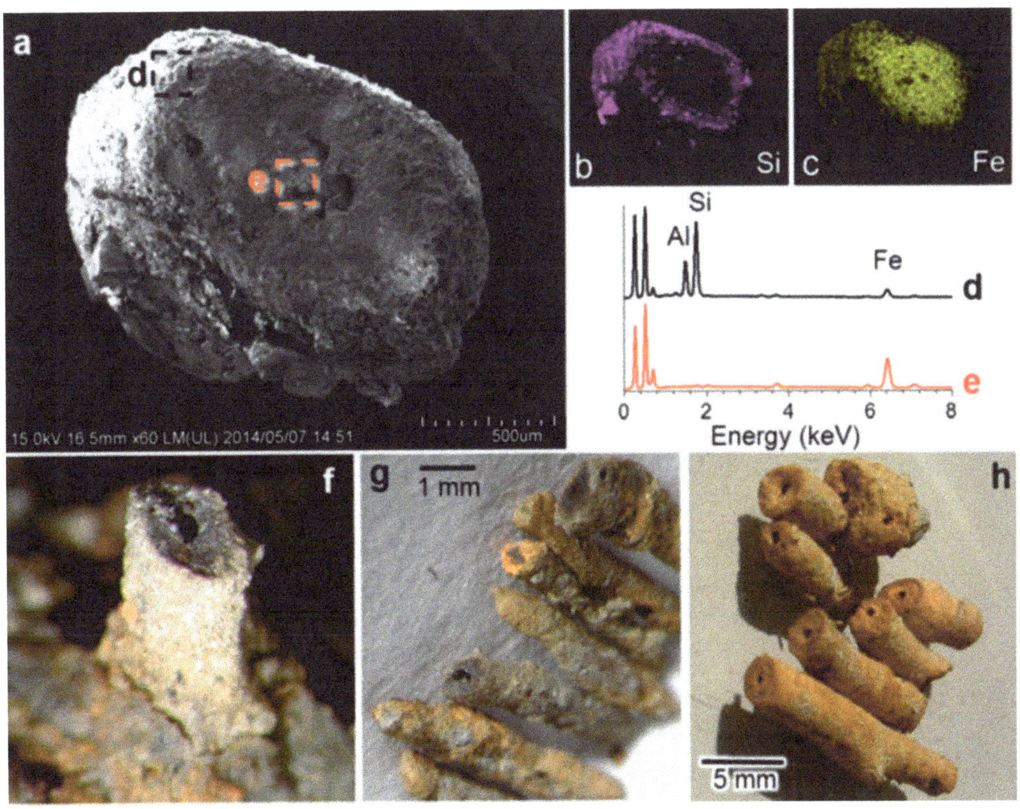

Fig. 5.23 Siderite resembling a filling pseudomorph separated from the soil profile shown in Fig. 5.22b at a depth of 40–50 cm. (**a**) SEM image, (**b** and **c**) Si and Fe element maps, respectively, (**d** and **e**) EDX spectra for the dashed areas (d) and (e), respectively, identified in (a), (**f**) an example of siderite having a central pore, (**g**) optical photographs of several rod-like siderites found at the same depth in Fig. 5.22b, (**h**) optical photograph of iron mottle-like root iron plaque found at a depth of 20–25 cm from the same soil profile as that shown in Fig. 5.5

the dashed square areas (d) and (e), respectively, shown in Fig. 5.23a. These results suggest that the outer part of the siderite rod mainly consists of aluminosilicates accompanied by hydrated iron oxide (Fig. 5.23d), and that the central part consists of siderite (Fig. 5.23e). The central pore was almost entirely filled with siderite. Figure 5.23f shows an optical micrograph of a similar rod. Although it has the central pore, the color of the cross-section is dominantly grayish. Several similar rods were separated from the same horizon, of which some were light brown and others were grayish in color (Fig. 5.23g). These rods appear to be formed from root iron plaque (Figs. 5.23h and 5.6), although the latter is more accurately cylindrical.

5.5 Pyrite and Related Sulfur-Containing Inorganic Constituents

Sulfur is an essential element of organisms. It is sensitive to redox reactions in soil and exists as both inorganic and organic constituents in soil. In this section, two sulfides and one sulfate are introduced. Although gypsum is also an important sulfate in soil, it is mainly introduced in Sect. 6.2.

5.5.1 Noncrystalline Iron(II) Sulfide

Dark-colored areas are sometimes found in submerged soil, particularly when sulfate-containing fertilizers are applied. Figure 5.24a shows rice cultivation with a large amount of gypsum in a glass container, such that roots can be observed occasionally, although the glass container was covered with a dark sheet to avoid growth of photosynthetic organisms on the inner surface. Figure 5.24b shows a close-up photograph of roots at the middle depth of the container. The thick roots are mostly brown, whereas thin lateral roots are mostly black. If there is no significant difference in organic matter supply to microorganisms between the thick and thin roots, the air supply to the thin roots is probably more restrictive than that to the thick roots (Trolldenier 1977). The deeply dark color of the fine roots suggests diffusion of iron and sulfur to the thin roots due to the low solubility of the precipitate. Figure 5.24c shows a cross-section of the soil with roots. The upper 4–5 cm is an oxidizing layer. Roots are distributed more along the glass container than in the bulk soil. Below the oxidizing layer, the color of both the reducing layer and roots in the soil is dark.

The dark-colored material is noncrystalline ferrous sulfide. After the roots were washed, they were freeze-dried. An optical micrograph (Fig. 5.25a) shows that the thick roots are dark brown, and the thin lateral roots are dark. EDX spectra obtained from the dashed outlined areas of a lateral thin root (Fig. 5.25c) and the thick root

Fig. 5.24 Rice cultivation in a glass container using 1.6 kg soil with N, P, K fertilizers and 8 g $CaSO_4 \cdot 2H_2O$. (**a**) soil and rice plants at harvest time, (**b**) close-up photograph of rice roots darkened by noncrystalline ferrous sulfide, (**c**) cross-section of soil removed from the glass container

(Fig. 5.25d) show that the dark material is noncrystalline ferrous sulfide. The S and Fe counts are much higher for the thin root than those for the thick root, in accordance with the intensity of dark color. However, a magnified SEM image (Fig. 5.25e) of the thin-root dashed area (Fig. 5.25c) shows only the root surface, possibly due to the very small particle size of noncrystalline ferrous sulfide (Fanning et al. 2002). The root surface appears as epidermis and/or exodermis from which cells were sloughed off. If the washed dark roots were left to stand in air, color of the noncrystalline iron sulfide easily disappeared in a day due to oxidation.

5.5.2 *Pyrite*

Pyrite is ferrous disulfide [FeS_2]. Noncrystalline ferrous sulfide may be a precursor of pyrite (Fanning et al. 2002). Pyrite may occur in newly reclaimed land from drainage or construction, and in the deep horizons of acid sulfate soil.

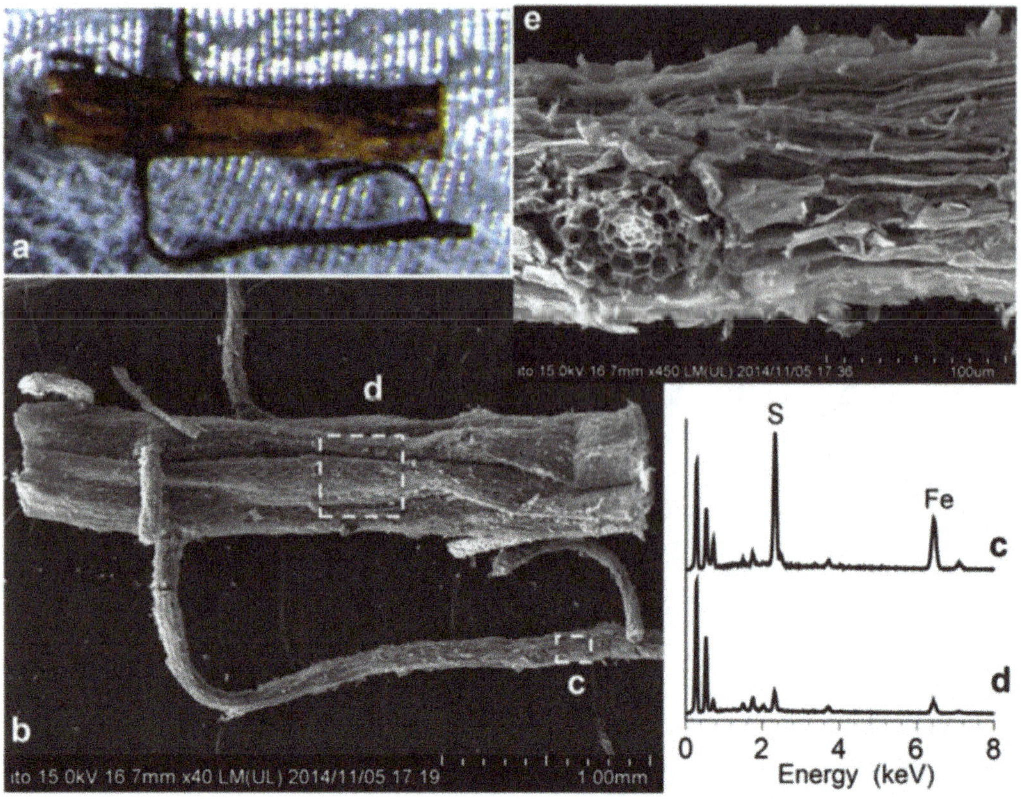

Fig. 5.25 Analyses of darkened rice root piece with gypsum application. (**a**) Optical photograph of rice root piece separated from the soil section shown in Fig. 5.24c, (**b**) SEM image of the rice root piece in (a), (**c** and **d**) EDX spectra for dashed areas (c) and (d) of the SEM image in (b), (**e**) magnified SEM image of dashed area (c) shown in (b)

Figures 5.26 and 5.27 show framboidal pyrite separated from the 2011 muddy tsunami deposit in Miyagi Prefecture, Japan (also see Sect. 6.2). At least at four sampling sites, muddy tsunami deposits with total S content between 8 and 23 g kg^{-1} contained framboidal pyrite, which is most commonly found in soils and sediments (Fanning et al. 2002). Figure 5.26a shows a SEM image of the framboidal pyrite. As shown in the EDX spectrum (Fig. 5.26b) and powder XRD pattern (Fig. 5.26c), and also in the element maps of S (Fig. 5.26d) and Fe (Fig. 5.26e), spherical or spheroidal crystal aggregates contain S and Fe as major elements; in addition, the powder XRD pattern includes a set of diffraction peaks from pyrite. Some of the crystal aggregates are partly broken. The pyrite samples shown in Figs. 5.26a and 5.27a were prepared after treatment by hot KOH digestion and separation of heavy mineral fractions (Valentyne 1963).

The crystal aggregates of pyrites are grayish gold or gold in color, and spherical or spheroidal in shape (Fig. 5.27a). Magnification of one of the crystal aggregates (Fig. 5.27b) shows that it consists of different crystal forms, such as octahedral (Fig. 5.27c) and polyhedral or sub-rounded (Fig. 5.27d) types.

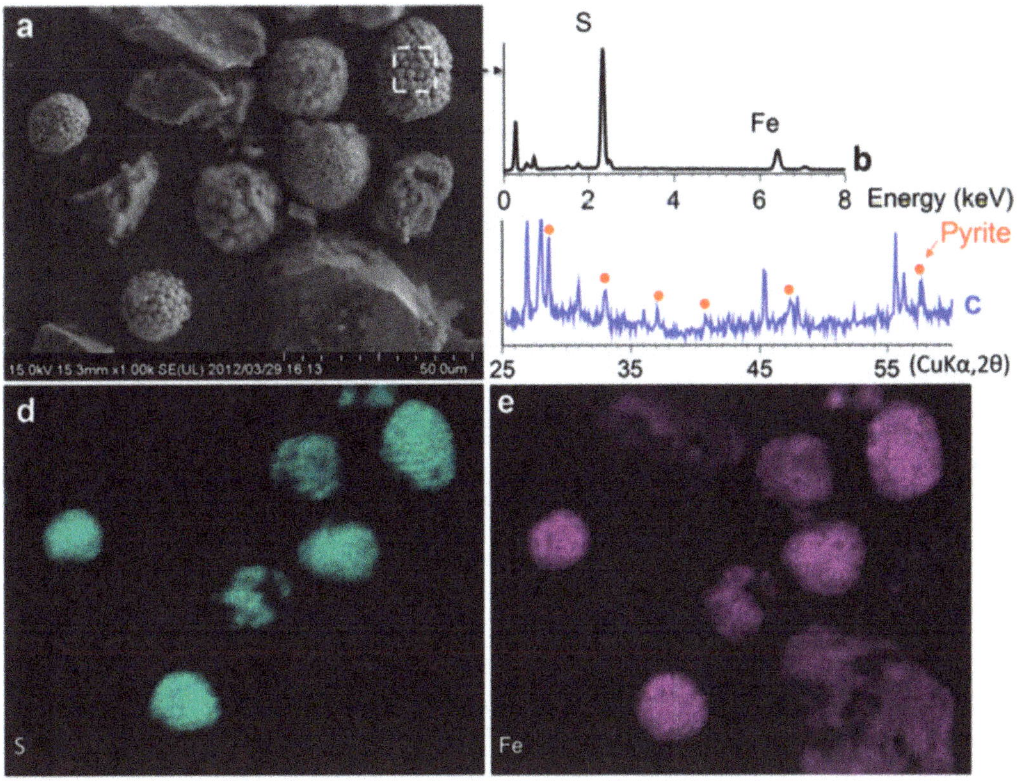

Fig. 5.26 Detection of framboidal pyrite. (**a**) SEM image including framboidal pyrite particles, (**b**) EDX spectrum of a framboidal pyrite particle, (**c**) XRD pattern including diffraction peaks from pyrite (red filled circles), (**d** and **e**) element maps for S and Fe, respectively

The presence of pyrite or sulfidic materials can be evaluated by pH measurement after H_2O_2 treatment or incubation of soil samples at field capacity for 8 weeks. If the pH after this treatment is lower than 3 or 3.5, the soil contains a significant amount of sulfidic materials. Sulfidic soils are converted to strongly acid soils after oxidation (Soil Survey Staff 1999).

5.5.3 *Jarosite*

Jarosite is a sulfate mineral [$KFe_3(SO_4)_2(OH)_6$] formed by oxidation of pyrite. A Na-substituted type of jarosite, natrojarosite [$NaFe_3(SO_4)_2(OH)_6$], also exists. They are both stable under acidic conditions. One of the characteristics of acid sulfate soil (Fanning and Burch 2000) is significant amounts of jarosite or natrojarosite. Acid sulfate soils are distributed in coastal lowlands, which are affected by sea-level rise and fall. Chemical problems of acid sulfate soil are low pH, lack of basic cations, excessive toxic Al, and high phosphate fixation. Soil amendments, such as liming

Fig. 5.27 Morphological properties of framboidal pyrite. (**a**) optical micrograph of framboidal pyrite particles, (**b**) magnified SEM image of a framboidal pyrite particle, (**c**) octahedral-type particles in framboidal pyrite, (**d**) polyhedral or rounded particles in framboidal pyrite

and application of other necessary fertilizers, are important in order to use these soils for agricultural production.

Figure 5.28 shows an example of natrojarosite. The sample field is presently used for sugarcane production (Fig. 5.28a). In the soil profile, the yellow patches at a depth of approximately 60 cm from the soil surface (Fig. 5.28b) are natrojarosite. The SEM image (Fig. 5.28c) shows pseudocubic crystal forms, and the EDX spectrum (Fig. 5.28d) obtained from the dashed square area (Fig. 5.28c) shows significant peaks for Na, S, and Fe. The powder XRD pattern (Fig. 5.28e) is similar to the reference pattern of natrojarosite (Fig. 5.28f).

Further readings for submerged soils and wetland soils are Kirk (2004), and Vepraskas and Craft (2016).

Fig. 5.28 An example of natrojarosite occurrence. (**a**) Landscape of soil containing sulfidic materials in the lower horizon, (**b**) soil profile containing natrojarosite (white arrow), (**c**) SEM image of natrojarosite, (**d**) EDX spectrum for the dashed area identified in (c), (**e**) XRD pattern of natrojarosite, (**f**) reference XRD pattern of natrojarosite (Brindley and Brown 1980)

References

Ando T, Yoshida S, Nishiyama I (1983) Nature of oxidizing power of rice plants. Plant Soil 72:51–57

Brindley GW, Brown G (1980) Crystal structures of clay minerals and their X-ray indentification, Mineralogical society monograph no.5. Mineralogical Society, London

Childs CW, Matsue N, Yoshinaga N (1991) Ferrihydrite in volcanic ash soils of Japan. Soil Sci Plant Nutr 37:299–311

Fanning DS, Burch SN (2000) Coastal acid sulfate soils, Reclamation of drastically disturbed lands, agronomy monograph, vol 41. American Society of Agronomy, Madison, pp 921–937

Fanning DS, Rabenhorst MC, Burch SN, Islam KR, Tangren SA (2002) Sulfides and sulfate. In: Dixon JB, Schulze DG (Co-eds) Soil mineralogy with environmental applications, Soil Science Society of America, Inc. Madison, pp 229–260

Fonseca EC, da Silva EF (1998) Application of selective extraction techniques in metal-bearing phases identification: a South European case study. J Geochem Explor 61:203–212

Fu Y-Q, Yang X-J, Ye Z-H (2016) Identification, separation and component analysis of reddish brown and non-reddish brown iron plaque on rice (Oryza sativa) root surface. Plant Soil 402:277–290

Garnd S, Lavkulich LM (1980) The oxidation mechanism of vivianite as studies by Moessbauer spectroscopy. Am Mineral 65:361–366

Global Rice Science Partnership (2013) Rice almanac, 4th edn. Los Banos (Philippines), International Rice Research Institute
Glossary of Soil Science Committee (2008) Glossary of soil science terms. Soil Science Society of America, Inc, Madison
Hansel CM, Fendorf S, Sutton S (2001) Characterization of Fe plaque and associated metals on the roots of mine-waste impacted aquatic plants. Environ Sci Technol 35:3863–3868
Heiberg L, Koch CB, Kjaergaard C, Jensen HS, Hansen HCB (2012) Vivianite precipitation and phosphate sorption following iron reduction in anoxic soils. J Environ Qual 40:938–949
Hurt GW, Whited PM, Pringle RF (eds) (1996) Field indentification of hydric soils in the United States. US Department of Agriculture, Natural Resources Conservation Service, Fort Worth
Ito J (1975) Concretion of ferrous phosphate ($Fe_3(PO_4)_2.8H_2O$) appears in grey horizon of paddy soils. Bull Hokuriku Natl Agric Exp Stn 18:119–14 (In Japanese, with English abstract.)
IUSS Working Group WRB (2015) World reference base for soil resources 2014. International soil classification system for naming soils and creating lengends for soil maps, update 2015, World soil resources reports no.106. FAO, Rome
Kahn N, Seshadri B, Bolan N, Saint CP, Kirkham MB, Chowdhury S, Yamaguchi N, Lee DY, Li G, Kunhikrishnan A, Qi F, Karunanithi R, Qiu R, Zhu Y-G, Syu CH (2016) Root iron plaque on wetland plants as a dynamic pool of nutrients and contaminants. Adv Agron 138:1–96
Kawai M, Samarajeewa PK, Barrero RA, Nshiguchi M, Uchiymiya H (1998) Cellular dissection of the degradation pattern of cortical cell death during aerenchyma formation of rice roots. Planta 204:277–287
Kirk G (2004) The biogeochemistry of submerged soils. John Wiley & Sons Ltd., Chichester 291p
Kojima M (1971) Micro-analysis of rusty mottlings in paddy soils. J Sci Soil Manure, Jpn 42:69–73
Komiyama T, Niizuma S, Fujisawa E, Morikuni (2013) Phosphorus compounds and their solubility in swine manure compost. Soil Sci Plant Nutr 59:419–426
Kondo M, Aguilar A, Abe J, Morita S (2000) Anatomy of nodal roots in tropical upland and lowland rice varieties. Plant Prod Sci 3:437–445
Kusunoki A, Nanzyo M, Kanno H, Takahashi T (2015) Effect of water management on the vivianite content of paddy-rice roots. Soil Sci Plant Nutr 61:910–916
Kyuma (2004) Paddy soil science. Kyoto University Press, Kyoto
Kyuma K, Mitsuchi M, Moormann FR (1988) Man-induced soil wetness: the “anthraquic” soil moisture regime. In: Kinloch DI, Shoji S, Beinroth FH, Eswaran H (eds) Proceedings of the ninth international soil classification workshop, Japan, 20 July to 1 August, 1987. Publ. by Japanese Committee for the 9th inernational soil classification workshop, for the soil management support service, Washinton DC, USA, pp 138–146
Lehr JR, Brown EH, Frazier AW, Smith JP, Thrasher RD (1967) Crystallographic properties of fertilizer compounds, Chemical Engineering Bull. 6. Tennessee Valley Authority, Knoxville
Lindsay WL (1979) Chemical equilibria in Soils. John Wiley & Sons, New York, pp 163–209
Nanzyo M, Kanno H, Obara S (2004) Effect of reducing conditions on P sorption of soils. Soil Sci Plant Nutr 50:1028–1028
Nanzyo M, Yaginuma H, Sasaki K, Ito K, Aikawa Y, Kanno H, Takahashi T (2010) Identification of vivianite formed on the root of paddy rice grown in pots. Soil Sci Plant Nutr 56:376–381
Nanzyo M, Onodera H, Hasegawa E, Ito K, Kanno H (2013) Formation and dissolution of vivianite in paddy field soil. Soil Sci Soc Am J 77:1452–1459
Ponnamperuma FN (1972) The chemistry of submerged soils. Adv Agron 24:29–96
Rothe M, Kleeberg A, Hupfer M (2016) The occurrence, identification and environmental relevance of vivianite in waterlogged soils and aquatic sediments. Earth-Sci Rev 158:51–64
Sadana US, Claassen N (1996) A simple method to study the oxidizing power of rice roots under submerged soil conditions. Z Pflanzenahr Bodenk 159:643–646
Sadiq M, Lindsay WL (1979) Selection of standard free energies of formation for use in soil chemistry. Tech Bull 134, Colorado State Univ Expt Stn
Saito M, Kawaguchi K (1971a) Flocculating tendency of paddy soils (Part 1) periodical changes of physical properties of paddy soils under flooded conditions. J Sci Soil Manure, Jpn 42:1–6
Saito M, Kawaguchi K (1971b) Flocculating tendency of paddy soils (Part 3) structure of poorly drained paddy soils. J Sci Soil Manure, Jpn 42:61–64

Schwertmann U (1973) Use of oxalate for Fe extractions from soils. Can J Soil Sci 53:244–246

Shiga H (1973) Effect of phosphorus fertility of soils and phosphate application on rice culture in cool region. Part 1. Measurement of phosphorus supplying ability of paddy soils. Res Bull Hokkaido Natl Agric Exp Stn 105:31–49 (In Japanese, with English abstract.)

Shiga H, Yamaguchi Y (1976) Effect of phosphorus fertility of soil and phosphate application on rice culture in cool region Part 3. On the relations with applied nitrogen and with variance of climate in continuous years. Res Bull Hokkaido Natl Agric Exp Stn 116:139–155 (In Japanese, with English abstract.)

Soil Survey Staff (1999) Soil taxonomy, a basic system of soil classification of making and interpreting soils surveys, USDA-NRCS, agriculture handbook no. 436, U.S. Government Printing Office, Washington, DC

Stumm W, Morgan JJ (1996) Oxydation and reduction; equilibria and microbial mediation. In: Aquatic chemistry, chemical equilibria and rates in natural waters, A Wiley-Interscience series of texts and monograph. Wiley Interscience, New York, pp 425–515

Togami K, Miura K, Ito K, Kanno H, Takahashi T, Nanzyo M (2017) Elemental affinity for siderite found in a Japanese paddy subsoil. Soil Sci Plant Nutr 63:101–109

Trolldenier G (1977) Mineral nutrition and reduction processes in the rhizosphere of rice. Plant Soil 47:193–202

Valentyne JR (1963) Isolation of pyrite spherules from recent sediments. Limnol Oceanogr 8:16–30

Vepraskas MJ (1992) Redoximorphic featrures for identifying aquic conditions. North Carolina State University Technical Bulletin 301, Raleigh, North Carolina, USA

Vepraskas MJ, Craft CB (2016) Wetland soils – genesis, hydrology, landscapes, and classification, 2nd edn. CRC Press, Taylor & Francis group, Boca Raton-London-New York

Wada H, Neue H–U (1988) Chemistry and biochemistry of paddy soils. In: Kinloch DI, Shoji S, Beinroth FH, Eswaran H (eds) Proceedings of the ninth international soil classification workshop, Japan, 20 July to 1 August, 1987. Publ. by Japanese Committee for the 9th inernational soil classification workshop, for the soil management support service, Washinton DC, USA, pp 115–12

Wada H, Miyashita K, Takai Y (1977) Studies on the decomposition process of plant residue in the paddy field. J Sci Soil Manure, Jpn 48:166–170 (in Japanese)

Walpersdorf E, Koch CB, Heiberg L, O'Connell DW, Kjaergaard C, Hansen HCB (2013) Does vivianite control phosphate solubility in anoxic meadow soils ? Geoderma 65:193–194

Yagi (1997) Methane emissions form paddy fields. Bull Natl Inst Agro-Environ 14:96–210

Yoshida H, Matsuoka K (2004) Occurrence of 'Takashikozo' from Takashihara area, Aichi-Prefecture. Bull Nagoya Univ Museum 20:25–34 (In Japanese with English abstract)

Zelibor JL Jr, Senftle FE, Reingardt JL (1988) A proposed mechanism for the formation of spherical vivianite crystal aggregates in sediment. Sediment Geol 59:125–142

Chapter 6
Role of Inorganic Soil Constituents in Selected Topics

Abstract Three topics are introduced to exemplify the important roles of inorganic soil constituents—the effects of tsunami on soil in Japan in 2011, the dynamics of radiocesium in the soil environment, and phosphates related to a soil–plant system. With respect to tsunami inundation into paddy field soils, soil erosion by seawater flow, sedimentation of soil transported by the seawater flow, precipitation of evaporites, and sodification are discussed. Removal of the deposited sediments and soil washing by rain and irrigation water were effective for restoration of the salt-affected farmlands. Radiocesium was effectively trapped by soil, which regulated its transfer to agricultural products. Among inorganic soil constituents, weathered biotite has a high fixation capacity for radiocesium. The biotite might have been released from granitic rock and volcanic ash. Apatite is the key phosphate in both natural and farmland soils, although it is converted to more soluble forms in the fertilizer industry. Fixation of phosphate by active Al materials is so high in Andisols that the recovery of phosphate by agricultural crops is low, and phosphate accumulation in plow layer soil is continuing. Struvite plays a role in cycling phosphate in the soil–plant system of farmlands.

6.1 Introduction

The inorganic constituents of soil play important roles in providing ecological services, as described in Chap. 1. The major inorganic constituents in soil were outlined in Chaps. 2, 3, 4 and 5. This final chapter is an attempt to exemplify the functioning of various soil inorganic constituents in different cases. Three topics are addressed in the chapter: the huge tsunami that occurred in northeastern Japan in 2011 including ion exchange reactions, the behavior of radiocesium in the soil environment including Cs^+ fixation, and reactions of phosphate with soil inorganic constituents. Some evaporites and phosphates that were not discussed in the previous chapters are introduced in relation to these topics.

M. Nanzyo, H. Kanno, *Inorganic Constituents in Soil*,
https://doi.org/10.1007/978-981-13-1214-4_6

6.2 Effects of Tsunami on Soils

"Tsunami" is a Japanese word that literally means harbor wave. A tsunami forms as the result of a conveyance of a sudden movement of the seafloor, an earthquake, to a huge amount of water, as shown schematically in Fig. 6.1a. The difference between a tsunami and a typical wave caused by strong wind or low atmospheric pressure is that the whole water column, from surface to bottom, moves in the former, whereas only surface water moves in the latter. Even if a tsunami is not very high in the deep ocean, it becomes large when it reaches a shallow harbor.

A magnitude 9.0 earthquake struck eastern Japan on March 11, 2011 (Norio et al. 2011). It triggered a huge tsunami, as shown in Fig. 6.1b, c. The area of farmland damaged by this tsunami is summarized in Fig. 6.2 and Table 6.1. Miyagi Prefecture was damaged most extensively. The tsunami penetrated as much as several kilometers inland from the coastline of Miyagi Prefecture. The damaged farmland consisted mainly of paddy fields (Table 6.1).

6.2.1 Survey and Analyses of the Tsunami–Affected Soils in Miyagi Prefecture

A wide area covering almost all of the farmland affected by the huge tsunami on the Pacific coast of Miyagi Prefecture was surveyed from May 11 to 19, 2011

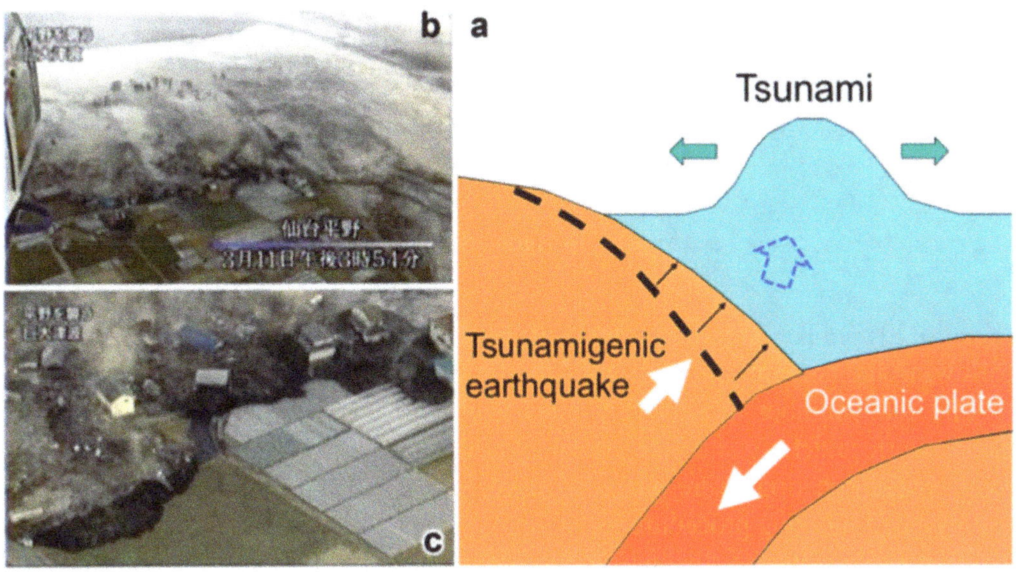

Fig. 6.1 Tsunami on March 11, 2011. (**a**) Schematic diagram showing the generation of the tsunami, (**b**) non-colored tsunami inundating land. (from NHK TV news), (**c**) change in the color of the tsunami when it fell down from the ridge. (From NHK TV news)

Fig. 6.2 Locations of prefectures where the Pacific coast was affected by the large 2011 tsunami

Table 6.1 Estimated areas affected the tsunami

Prefecture	ha	Paddy field ha	Upland ha
Aomori	79	76	3
Iwate	1,838	1,172	666
Miyagi	15,002	12,685	2,317
Fukushima	5,923	5,588	335
Ibaraki	531	525	6
Chiba	227	105	122
Total	23,600	20,151	3,449

Ministry of Agriculture, Fishery and Forestry, Japan (2012)

(Kanno 2017). The area had a total of 70–100 mm of rain during the 2 months following the tsunami inundation of the farmland.

The total number of sampling sites was 344 fields chosen mostly from the tsunami-affected farmland. For each field, samples taken from the same layer at two separate locations were mixed to represent the field. Based on the preliminary survey, the following procedure for sampling tsunami deposits and underlying soils was determined. If the thickness of the tsunami deposit was 1 cm or more, the tsunami deposit was sampled separately from the underlying original soil. If the tsunami deposit could be separated into sandy and muddy deposits, these deposits were sampled separately. The original soil was sampled from two layers, 0 to 10 cm (original soil 1) and 10 to 20 cm (original soil 2) from the boundary between the tsunami deposit and the top of the original soil. The boundary between the tsunami deposit and the original soil was evident because the bottom of the tsunami deposit had a coarse texture, because sands settle faster than finer particles in a soil suspension, and the top of the original soil had a comparatively finer texture (Fig. 6.4c).

After air-drying the tsunami deposit and original soil samples, the samples were ground gently in a porcelain mortar and passed through a 2 mm sieve to prepare a fine earth fraction. As a result, the samples were mostly passed through the 2 mm sieve. Using these fine-earth fractions, total C, N, and S content, pH(H_2O) and pH (KCl) at the soil:water ratio of 1:2.5, electric conductivity at the soil:water ratio of 1:5 (EC(1:5)), the contents of water-soluble Na, K, Ca, and Mg in the 1:5 water suspension, and NH_4-exchangeable Na, K, Ca, and Mg were determined. The contents of NH_4-exchangeable Na, K, Ca, and Mg were determined by extracting the sample obtained after extraction of water-soluble Na, K, Ca, and Mg with 1 mol L^{-1} NH_4 acetate (pH 7) twice (Thomas 1982). Entrained solution volume after centrifugation and decantation of water extract was obtained by weight measurement, and the amount of each cation in the entrained solution was subtracted from each total cations found in the NH_4 acetate-extract for the calculation of NH_4 exchangeable cations.

The tsunami caused severe damage to buildings. However, the physical damage to farmland was not very severe compared to that of buildings, probably because the tsunami affected only the surface of the farmland and a direct hit between tsunami and farmland soil was mostly avoided.

Intensively eroded sites were found on the inland side of the road running along the coast (Fig. 6.3a). The distance between the road and the shoreline was less than 1 km. When the tsunami dropped from the slightly higher road, paddy field soil was excavated and removed. At these sites, the soil removal was greater than several tens of centimeters from the original surface of the paddy fields. Similar soil erosion occurred along ridges (30–40 cm high), as shown in Fig. 6.3b, although the intensity of the erosion at these sites was lower than that along the roads and the plow sole mostly remained. The intensity of soil erosion over the wide surface of farmland was dependent on whether or not the farmland had been plowed. If the soil had not been plowed, erosion was limited and the rice stubble after harvest mostly remained (Nanzyo 2012). In contrast, plowed soil was removed by the tsunami, especially in farmland near the coast. Soil removal was lower in the farmland distant from the coast, even if the land had been plowed.

These interactions between tsunami and farmland are summarized in Fig. 6.3c. Erosion occurred at sites ①, ②, and ③, and deposition occurred at sites ④ and ⑤. Ion exchange and precipitation reactions occurred as chemical interactions at sites ④ and ⑤ (Fig. 6.3c). If there was a muddy (sometimes containing sulfides) and/or sandy deposit beneath the shallow seawater or in the nearshore zone (①) including the Teizan canal, the deposits might have been transported to farmland and deposited at sites ④ and/or ⑤. Moreover, the A_p horizon soil, after tilling, was at least partly lost. Thus, the deposits on the farmland also contain the eroded A_p horizon soil. The chemical reactions at sites ④ and ⑤ include the exchange reaction between Na^+ in

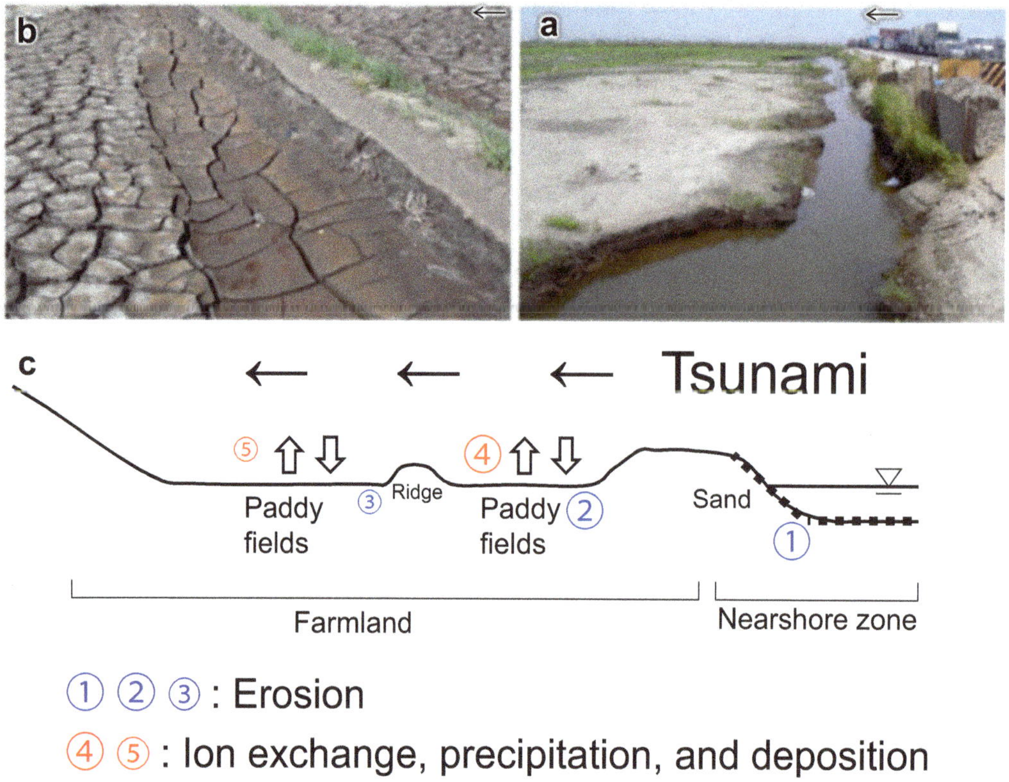

Fig. 6.3 Effects of the tsunami on farmland. (**a** and **b**) Erosion and deposition along a road and ridge, respectively, (**c**) schematic diagram of interactions between tsunami and farmland. Arrows in (**a**) and (**b**) show the direction of the tsunami

seawater and exchangeable cations in the A_p horizon soil and also the precipitation of $CaSO_4 \cdot 2H_2O$ and NaCl as the soils became dry.

Figure 6.4a, b show the distribution of the total thickness of the tsunami deposit (mud plus sand) in Miyagi Prefecture and a close-up illustration of the Sendai Bay area, respectively. The total thickness of the tsunami deposit tended to be thicker in the area near the shoreline. The total thickness was very thin or zero near the inland end of the tsunami-affected area. The sandy tsunami deposit was thicker near the shore than in the inland area. The muddy deposit tended to be thick in the intermediate area. Separation of the sandy and muddy deposits is somewhat clear (Fig. 6.4c), possibly because the suspension remained in the paddy field, where the ridge was 30–40 cm high (Fig. 6.3b).

There was a concern that abundances of toxic elements such as Cd, Cu, and As might be excessive in the tsunami deposits. However, according to Miyagi Prefecture, the concentrations of these elements were estimated to be lower than the upper limit established by law for cultivated soils in Japan, except that one mud sample exceeded the limit for As. However, because the thickness of this mud layer was 1 cm and the exceedance was small, the problem was not considered to be very severe (Shima et al. 2012; Inao et al. 2013).

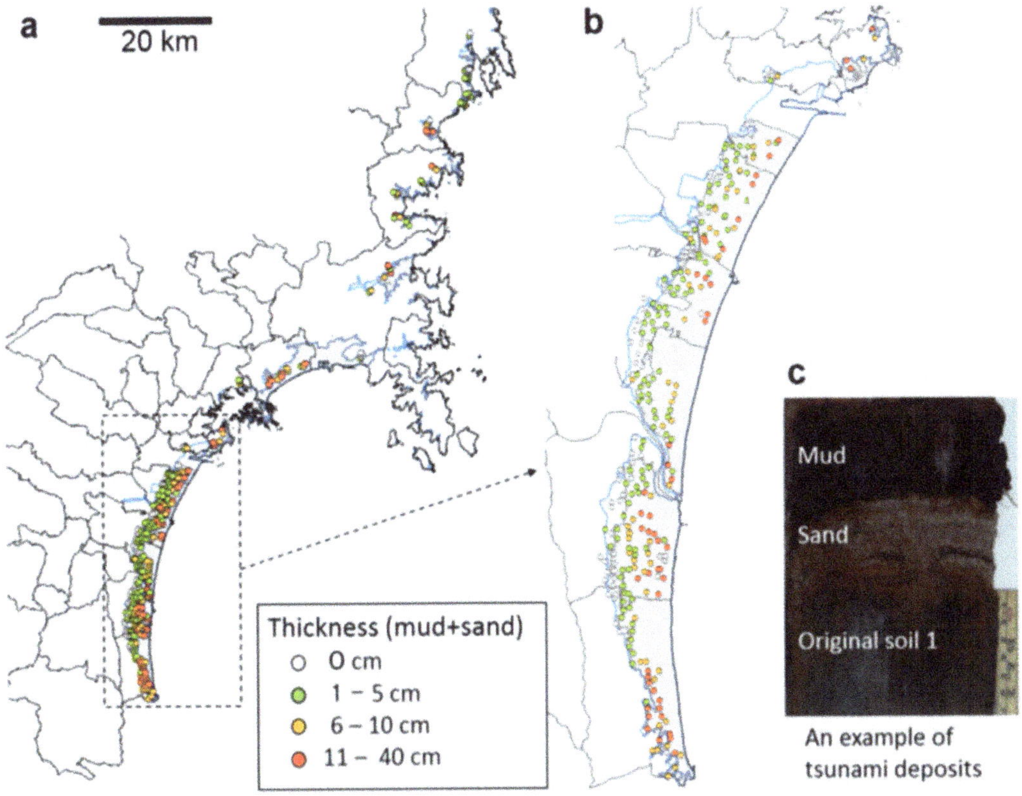

Fig. 6.4 Thickness of the tsunami deposit. (**a**) Total thickness of mud plus sand in Miyagi Prefecture, (**b**) close-up illustration of the Sendai Bay area, (**c**) an example of the tsunami deposit. (Shima et al. 2012; Inao et al. 2013)

Analytical results of the tsunami deposits and the underlying original soils are summarized in Table 6.2. The EC(1:5) values rose with inundation by seawater. Although the highest EC(1:5) values were observed in the muddy tsunami deposit, the original soils showed higher EC(1:5) values than the ordinary soils, indicating that the effects of seawater reached most of the original soil 2. The total C, N, and S content of the muddy tsunami sediment was higher than the content in the sandy tsunami deposits and original soils. Water-soluble and exchangeable Ca^{2+}, Mg^{2+}, K^{+}, and Na^{+} content was also high in the muddy tsunami deposit due to the effect of seawater. The EC(1:5) values were highly correlated with the sum of water-soluble $Ca^{2+} + Mg^{2+} + K^{+} + Na^{+}$ on a positive charge basis. Among the four exchangeable cations, the exchangeable sodium percentage (ESP) is used as an index of sodification. In this monograph, the value of 100 times the exchangeable Na divided by the sum of exchangeable $Ca^{2+} + Mg^{2+} + K^{+} + Na^{+}$ on a positive charge basis is used as ESP. The seawater also affected the pH(H_2O) values. The mean pH(H_2O) values were higher in the muddy and sandy tsunami deposits than in the original soils. Truog P values were higher in the muddy tsunami deposit than in the other deposits. A considerable sulfide concentration in some of the muddy tsunami deposits was suggested from their pH(H_2O_2) value of 3 or less (Shima et al. 2012). These results are discussed further below.

Table 6.2 Analytical results of tsunami deposits and underlying original soils of the tsunami-affected farmland in Miyagi Prefecture, Japan

	Tsunami deposit						Original soil					
	Muddy (229)[a]			Sandy (161)			1 (344)			2 (340)		
	Maximum	Mean	Minimum	Maximum	Mean	Minimum	Maximum	Mean	Minimum	Maximum	Mean	Minimum
Total (g kg^{-1})												
organic C	104	40.7	1.06	63.3	12.0	0.64	112	25.0	1.01	204	24.8	1.81
N	7.32	3.10	nd	4.79	0.77	nd	7.24	1.91	nd	11.7	1.78	nd
S	23.2	4.14	nd	3.60	0.59	nd	6.23	0.76	nd	2.87	0.66	nd
EC(1:5) (dS m^{-1})	56.1	15.1	0.19	19.1	2.31	0.08	11.9	2.36	0.06	4.80	1.31	0.06
Water-soluble ($cmol_c$ kg^{-1})												
Ca^{2+}	48.3	10.6	0.07	13.8	1.13	0.01	6.46	1.05	0.01	5.34	0.97	0.01
Mg^{2+}	59.8	13.7	0.02	18.2	1.34	0.01	11.7	1.08	0.01	2.90	0.57	0.01
K^+	3.90	1.41	0.03	1.57	0.24	0.02	1.20	0.22	0.02	0.65	0.10	0.01
Na^+	239	62.2	0.58	81.9	8.84	0.32	44.3	8.75	0.12	17.6	4.05	0.08
Sum	351	88.0	1.10	115	11.6	0.36	63.7	11.1	0.30	22.8	5.70	0.31
Exchangeable ($cmol_c$ kg^{-1})												
Ca^{2+}	40.0	10.19	1.05	22.9	7.15	0.50	25.0	5.90	0.25	25.5	7.71	0.21
Mg^{2+}	15.2	7.53	0.91	9.26	2.45	0.45	7.69	3.22	0.36	7.18	2.77	0.11
K^+	4.79	1.84	0.22	2.83	0.58	0.08	3.38	0.60	0.09	2.90	0.43	0.04
Na^+	16.9	5.74	0.26	7.69	1.55	0.06	6.29	2.12	0.05	4.40	1.32	0.03
Sum	64.3	25.3	2.79	34.1	11.7	1.73	28.6	11.8	1.15	28.5	12.2	0.51
pH(H_2O)	8.9	6.2	3.5	9.2	6.8	4.6	8.6	5.2	4.2	8.0	5.3	3.4
pH(KCl)	7.9	5.8	3.2	8.2	6.3	4.1	7.5	4.5	3.6	7.2	4.7	3.0
pH(H_2O)-pH(KCl)	1.4	0.34	0.05	1.9	0.53	0.05	2.0	0.64	0.16	1.8	0.67	0.19
Truog P (g P_2O_5 kg^{-1})	1.72	0.349	0.024	0.624	0.156	0.039	2.65	0.248	0.027	1.72	0.150	0.003

nd not detected due to the value being lower than the detection limit

[a]Parentheses show the number of each sample type

6.2.2 Origin of the Muddy Tsunami Deposit

In general, erosion by the tsunami occurred in the nearshore zone (Fig. 6.3 ①). (Srisutam and Wagner 2010). The color of a tsunami may change from place to place depending on the properties of the sediments on the seafloor. As shown in Fig. 6.1b, the tsunami that struck the Sendai Plain had a whitish color, suggesting that its mud content was low. The color of the tsunami became black when the tsunami crossed over the ridges and flowed down to the paddy fields (Figs. 6.1c and 6.3b). Eroded sites were also found along roads, where the tsunami crossed over the roads and flowed down to the paddy fields (Fig. 6.3a).

The distribution of the total thickness of the tsunami deposit is shown in Fig. 6.4. The sandy tsunami deposit was thicker near the shore, suggesting that most of the sandy tsunami deposit originated in the nearshore zone (Fig. 6.3c ①). However, the muddy tsunami deposit was thick in the intermediate area of tsunami inundation, so the origin of the muddy tsunami deposit was uncertain. From the distribution of total organic carbon (TOC) content of the muddy deposit shown in Fig. 6.5a, three areas having TOC higher than 4% can be identified (closed circles). A similar pattern of TOC distribution was found in the original soil 1 (Fig. 6.5b). Furthermore, the soil map of this area showed that TOC-rich soils, such as peat soil and muck soil, are

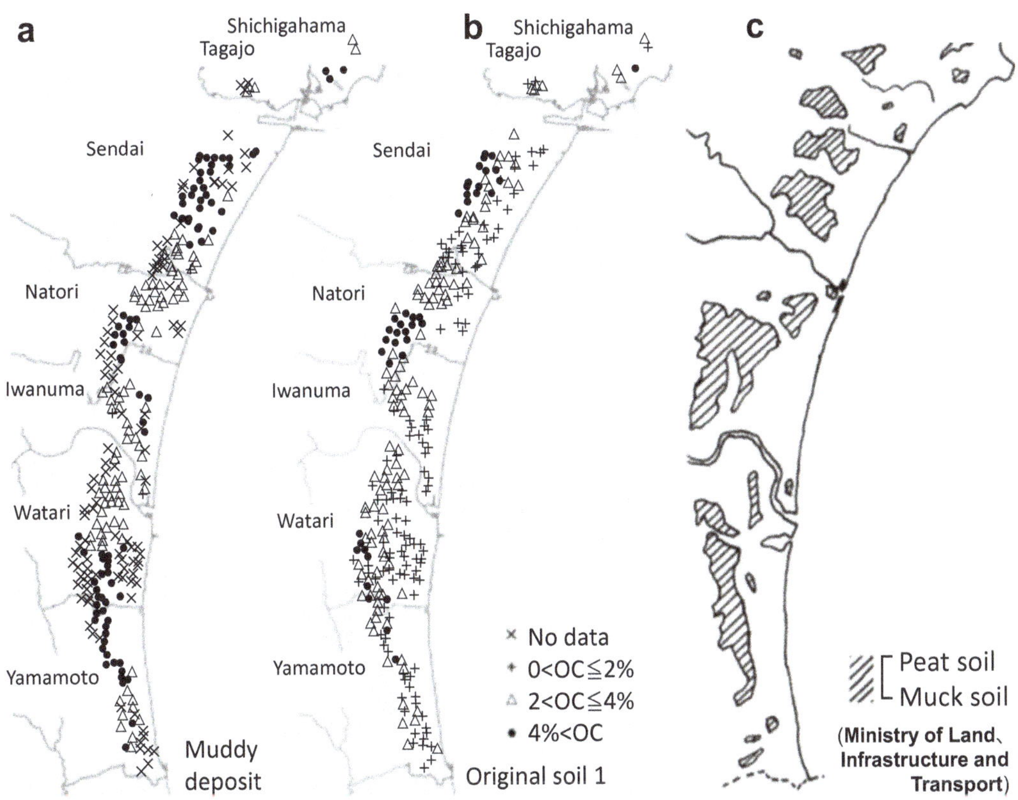

Fig. 6.5 Distribution of total organic carbon (TOC) concentration. (**a**) Muddy tsunami deposit, (**b**) underlying original soil 1, (**c**) soil map showing the distribution of TOC-rich soils (peat and muck soils). The reason for "No data" is that the thickness of the muddy tsunami deposit is very thin

distributed within the region (shaded areas in Fig. 6.5c). Hence, the TOC-rich areas of the muddy tsunami deposits are nearly identical to those of the original soil 1 and the soil map made before tsunami inundation. Considering these results, it is highly probable that the muddy tsunami deposits originated from the surface soil of the tsunami affected areas. This estimation is supported by the fact that the Truog P values of the muddy tsunami deposits are distinctively higher than those of the sandy tsunami deposit and original soils 1 and 2 (Table 6.2). The high Truog P values are due to the accumulation of P fertilizer in the plow layer soil. The plow layer soil of the farmland eroded by the tsunami was deposited with sorting as the tsunami retreated. Szczuciński et al. (2012) reported that marine sediments are present in only low abundance in the present muddy tsunami deposit on the Sendai Plain.

6.2.3 *Relationships Between TOC, TN, and TS of the Tsunami Deposits and the Original Soils*

The abundance of sulfides in the tsunami deposits was one of the concerns about the deposits. Total sulfur (TS) content was determined in addition to TOC and total nitrogen (TN) because acidification can occur after oxidation. The relationship between TOC and TS is plotted with TN in Fig. 6.6. In general, there is a close relationship between TOC and TN for both the muddy tsunami deposit and the original soil 2. A linear relationship was also found between the TOC and TS content of the original soil 2, suggesting that a major portion of the sulfur is in organic forms

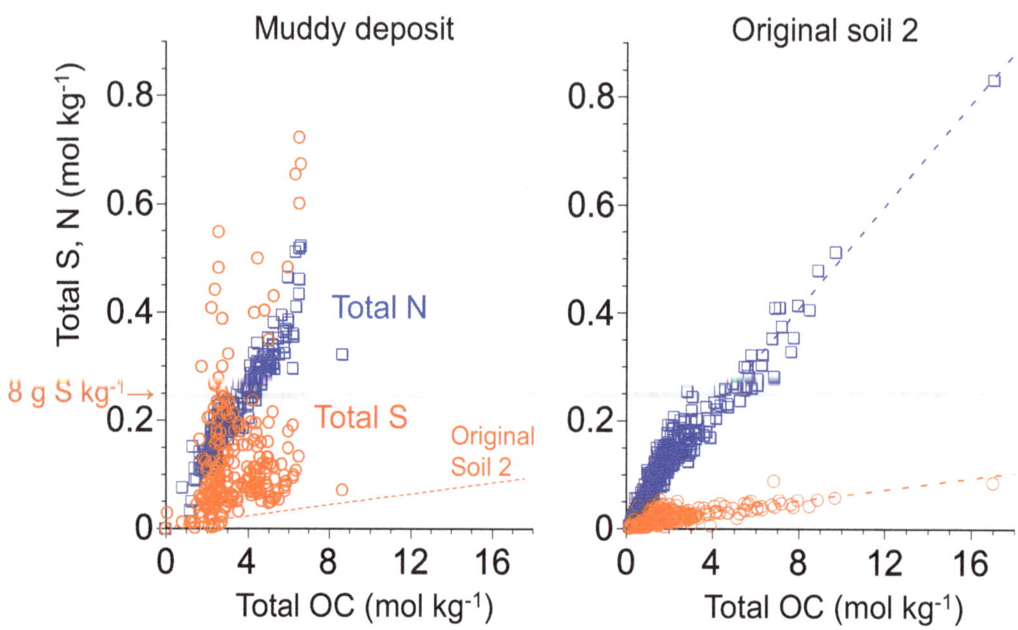

Fig. 6.6 Relationships between total organic C content and total N and S content of the muddy tsunami deposit and original soil 2, respectively

(Erickson 2009). In contrast, the TS content of the muddy tsunami deposit was larger than the content of original soil 2 by various amounts. Because the TS content of the muddy tsunami deposit does not correlate with TOC, the added TS (beyond that of original soil 2) appears to be in inorganic form. The four samples from TS-rich muddy tsunami deposits all included framboidal pyrite, as shown in Figs. 5.26 and 5.27.

6.2.4 *Evaporites on the Tsunami Deposits*

White powdery materials, possibly evaporites, formed on the surface of the muddy tsunami deposits, as shown in Fig. 6.3b. A sample was taken from one muddy tsunami deposit (Fig. 6.7a) and was examined using SEM-EDX and XRD. Halite and gypsum were identified, as shown in Fig. 6.7. The halite particles showed cubic form (Fig. 6.7b) and the gypsum crystals were prismatic (Fig. 6.7c) under the SEM. The elemental compositions of the halite and the gypsum were confirmed by EDX spectra, as shown in Fig. 6.7d, e, respectively. According to the XRD data, gypsum

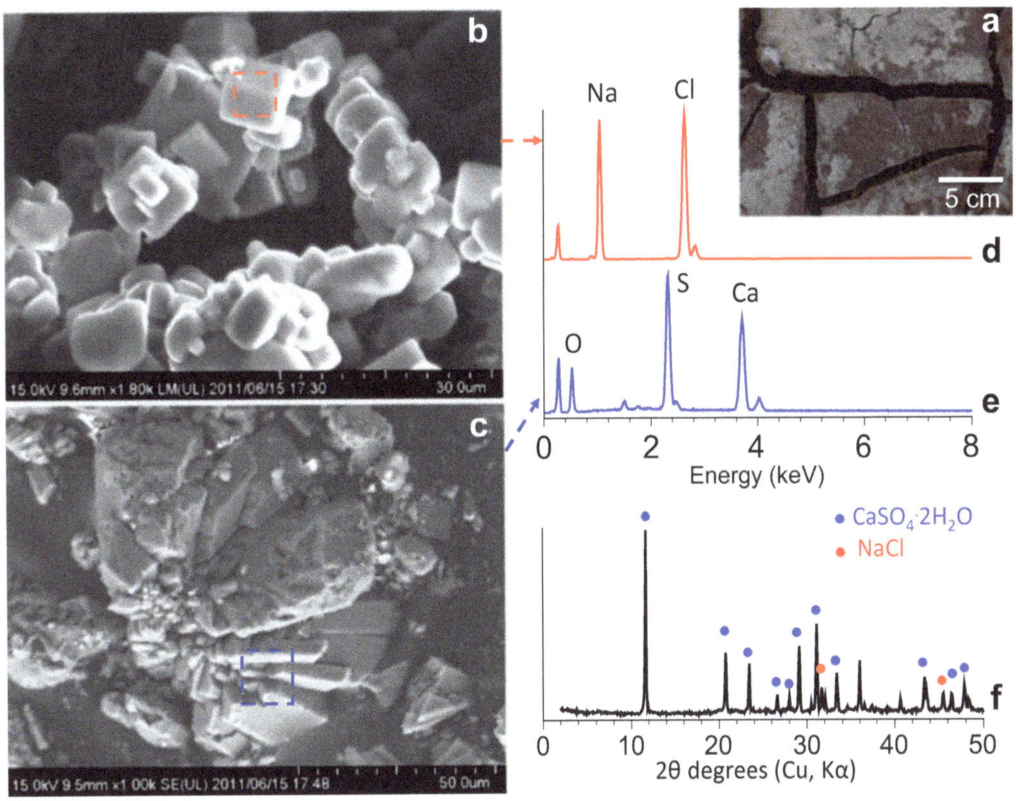

Fig. 6.7 Evaporites formed on the surface of the muddy tsunami deposit. (**a**) Optical photograph of the muddy tsunami deposit having evaporites, (**b** and **c**) SEM images of halite and gypsum, respectively, (**d** and **e**) EDX spectra of the dashed squares in (**b**) and (**c**), respectively, (**f**) XRD pattern of the evaporites

appears to be the major constituent in this sample (Fig. 6.7f). This result is due to the lower solubility of gypsum than halite. At the sampled site, a considerable portion of the halite had been removed by rainwater, leaving the gypsum as the major evaporite. Gypsum can precipitate from seawater during drying. In addition, Ca^{2+}, released from the Ap horizon soil by an ion exchange reaction with Na^+, could take part in the gypsum precipitation because SO_4^{2-} is relatively abundant in seawater.

Further study for evaporates is Doner and Lynn (1989).

6.2.5 Salinization and Sodification

The major chemical effects of tsunami inundation are salinization and sodification of soils (Fig. 6.8). Salinization is an increase in the concentration of water-soluble salts to a level detrimental to mesophytes or most crop plants. Saline soil is defined by an EC value of 4 dS m^{-1} using saturation extract. It roughly corresponds to an EC(1:5) value of 0.6 dS m^{-1}. Sodic soil is defined by an ESP value of 15% or more. Soils having both saline and sodic properties are saline-sodic soils. Typical farmland soils have exchangeable and water-soluble ionic compositions of $Ca^{2+} > Mg^{2+} > K^+ \simeq Na^+$ on a charge basis. The occurrence of a tsunami changes this ionic composition. With an increase in ESP, clay particles tend to disperse in water and become hard clods after drying.

According to the mean EC(1:5) values (Table 6.2), the extent of salinization was in the order of muddy tsunami deposits > sandy tsunami deposits $\simeq$ original soils 1 > original soils 2. The mean ESP value was highest in the muddy tsunami deposits and lowest in the original soils 2 among the four groups of samples. Changes in the charge fractions of exchangeable Ca^{2+}, Mg^{2+}, and K^+ compared with that of Na^+ for original soil 2, which showed a wide range of salinization and sodification, are plotted in Fig. 6.9. An increase in the charge fraction of exchangeable Na^+ indicates an increasing effect of seawater. The charge fraction of Ca^{2+} deceased with an increase of charge fraction of Na^+, and the charge fractions of Mg^{2+} and K^+ increased slightly with that of Na^+. These results are similar to those for the muddy and sandy tsunami deposits and original soil 1, although the slopes of the

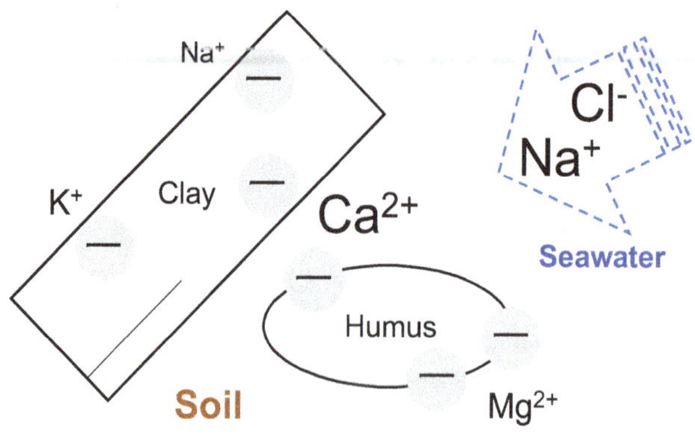

Fig. 6.8 Schematic diagram showing seawater inundation to soil

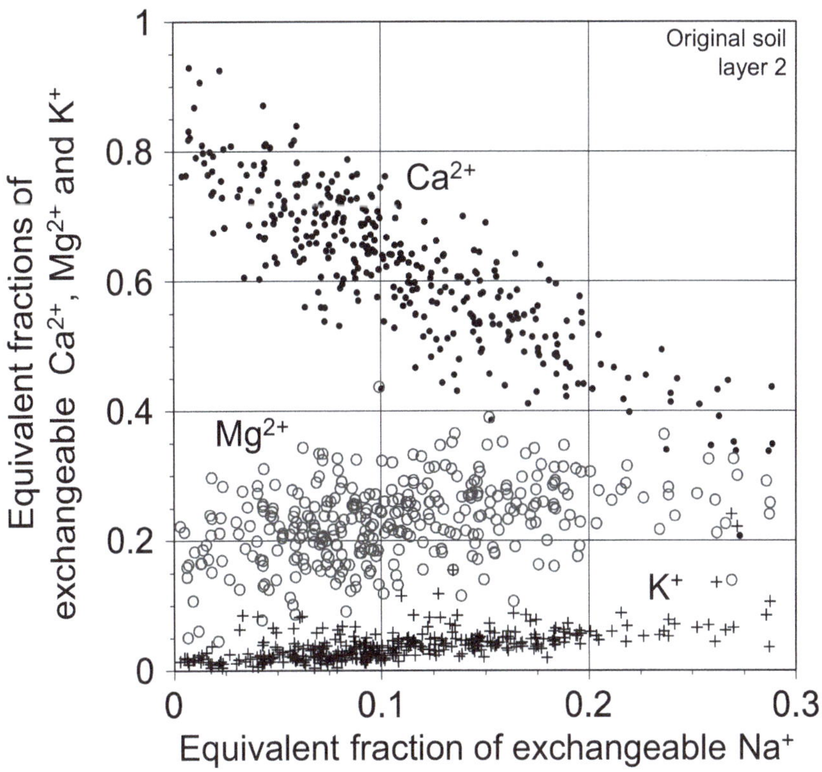

Fig. 6.9 Changes in charge fractions of exchangeable Ca^{2+} (•), Mg^{2+} (○), and K^+ (+) with exchangeable Na^+, of original soil 2

plot distributions are slightly diverse. The changes in the charge fractions of exchangeable cations are related to those ions in the solution phase in Figs. 6.10 and 6.11.

The relationships of the charge fractions of the four ions (Ca^{2+}, Mg^{2+}, K^+, and Na^+) between water-soluble (1:5 water extract) and exchangeable forms are plotted in Fig. 6.10. The charge fraction of exchangeable Na^+ increased gradually with an increase in the charge fraction of water-soluble Na^+, whereas the charge fraction of exchangeable Ca^{2+} decreased steeply with a decrease in the charge fraction of water-soluble Ca^{2+}. Although the distribution ranges of the charge fractions of K^+ and Mg^{2+} are narrower than those of Na^+ and Ca^{2+}, the charge fractions of exchangeable K^+ and Mg^{2+} increased with increases in the charge fractions of water-soluble K^+ and Mg^{2+}. These cation exchange reactions suggest that the high charge fraction of exchangeable Ca^{2+} for soils weakly affected by seawater is due to the high charge fraction of water-soluble Ca^{2+}.

The relationship between the water-soluble and exchangeable cations of the muddy tsunami deposit (Fig. 6.11) lacks high charge fraction plots of both water-soluble and exchangeable Ca^{2+} and low charge fraction plots of both water-soluble and exchangeable Na^+ in comparison with those for original soil 2 (Fig. 6.10). In Fig. 6.11, under the high charge fraction of water-soluble Na^+, the charge fractions of exchangeable Ca^{2+} and Mg^{2+} still maintain slightly higher levels than those of exchangeable Na^+. Plots similar to those of Fig. 6.11 were obtained from the sandy

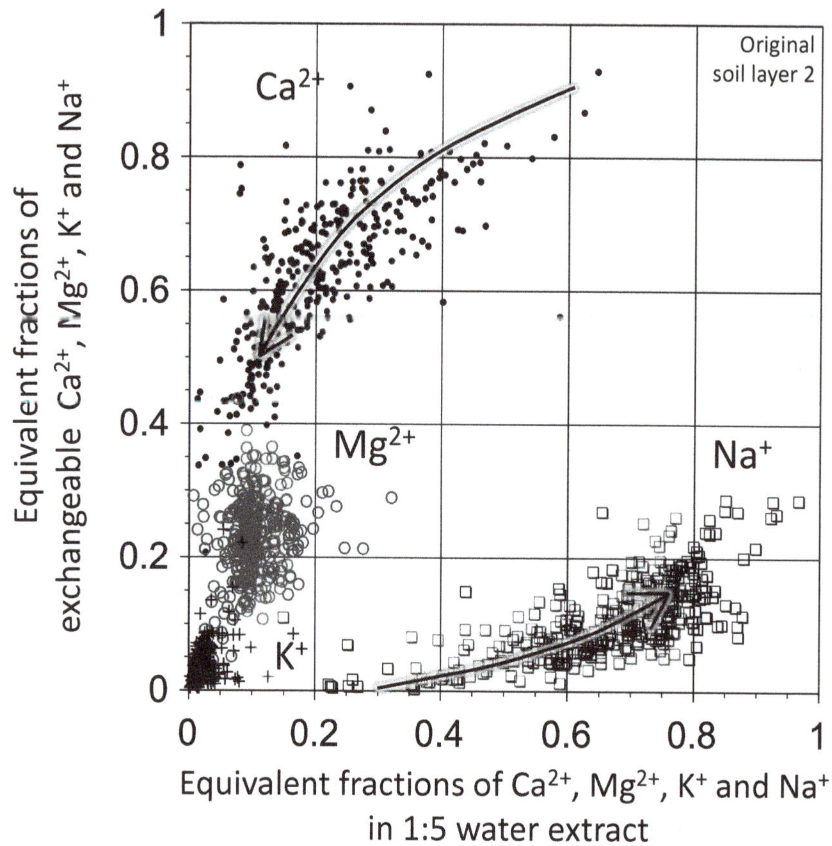

Fig. 6.10 Relationships between the charge fractions of water-soluble and exchangeable Ca^{2+} (•), Mg^{2+} (○), K^{+} (+), and Na^{+} (□), respectively, of original soil 2

tsunami deposit, and plots intermediate between those of Figs. 6.10 and 6.11 were obtained from original soil 1. Although the details of the high TS content (percentages of pyrite, gypsum, and organic forms) of the muddy tsunami deposit were not determined, dissolution of gypsum might have affected the water-soluble and exchangeable Ca^{2+} content to some extent. The charge fraction of exchangeable Na^{+} hardly rises to 0.5 or more under these conditions.

The cation exchange properties observed in the present tsunami-affected deposits and soils were compared with those of previous observation. The U.S. Salinity Laboratory used the following equation for a number of different soils and reported that the K_G was 0.01475 $(\text{mmol L}^{-1})^{-0.5}$ (Kamphorst and Bolt 1976).

$$\text{ESP}/(100 - \text{ESP}) = K_G \times \text{SAR} \tag{6.1}$$

SAR (sodium absorption ratio) = Water-soluble Na^{+}/(Water-soluble Ca^{2+} + $Mg^{2+})^{0.5}$

In the above, water-soluble Na^{+}, Ca^{2+}, and Mg^{2+} are expressed as (mmol L^{-1}) in 1:5 water extraction.

The relationship between the left side and the SAR of Eq. 6.1 is examined in Fig. 6.12. The K_G value of 0.015 is also shown in Fig. 6.12 by a dashed line. Although the scattering of plots is larger than the scattering reported by the

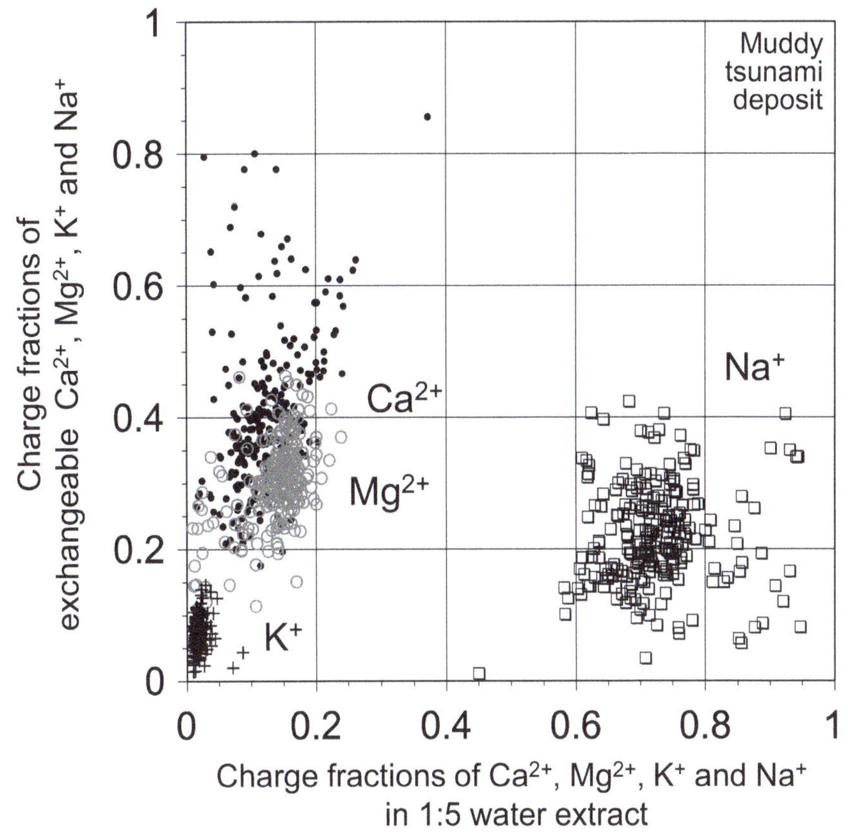

Fig. 6.11 Relationships between the charge fractions of water-soluble and exchangeable Ca^{2+} (•), Mg^{2+} (○), K^{+} (+), and Na^{+} (□), respectively, of the muddy tsunami deposit

U.S. Salinity Laboratory (1954), the plots are distributed around the K_G value of 0.015. Possible reasons for deviation of the plots from $K_G = 0.015$ are (i) the effect of gypsum dissolution on the water-soluble Ca^{2+} and exchangeable Ca^{2+} of the muddy tsunami deposit and (ii) the widely different soluble cation concentration expressed by the broad range of EC(1:5) values (Table 6.2). The clay mineral composition of the area is a mixture of smectite, vermiculite, and kaolin minerals. As widely different K_G values have been reported for different pure clay minerals, the almost constant values for many soils might be due to mixed clay mineral compositions (Shainberg et al. 1980; Miller et al. 1990; Kopittke et al. 2006; Endo et al. 2002). Because the average SAR of the river water used for irrigation of the tsunami-affected area is 0.6 (mmol $L^{-1})^{0.5}$ and the calculated ESP value using Eq. (6.1) is 0.9%, it is suggested that the ESP values of the tsunami-affected soils will decrease gradually with irrigation by the river water.

Further readings about the cation exchange reactions in soil include Evangelow and Phillips (2005) and McBride (1989).

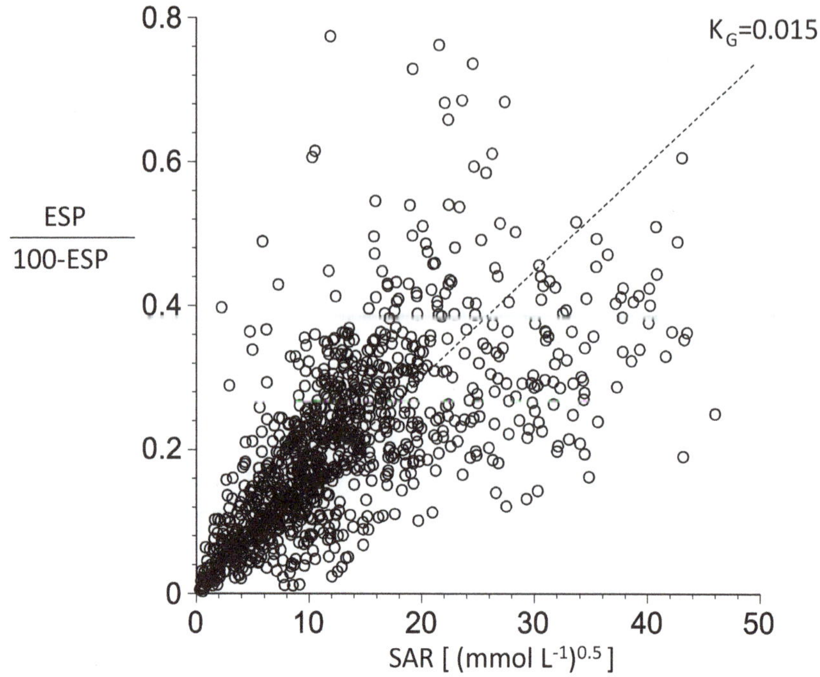

Fig. 6.12 Relationship between SAR and ESP/(100 − ESP) of the tsunami deposits and original soils 1 and 2. The dashed line shows the K_G value of 0.015. The number of samples is 1074

6.2.6 Variation in pH of Tsunami Deposits and Original Soils

Soil pH is one of the important variables which relate with many other soil properties. The pH(H_2O) values of the air-dried fine-earth fraction of paddy field soils in Japan typically ranges between 5.4 and 5.9 (Oda et al. 1987). The pH value of negatively charged soil decreases with the addition of salts, as exemplified by pH (KCl). As shown in Fig. 6.13a, the frequency distributions of pH(H_2O) values of the tsunami deposits and original soils are different, despite the muddy tsunami deposit being derived mostly from the farmland. The tsunami deposits were suspended once in seawater, which has a salt concentration of approximately 0.6 mol L^{-1}. The wide distribution of the pH(H_2O) values is due to differences in subsequent drainage conditions. Under well-drained conditions, the tsunami deposits were effectively washed by rainwater and the pH(H_2O) value became high. This is mainly due to the coordination of H^+ on the negative variable-charge site with a decrease in the electrolyte concentration in the liquid phase. As a result, OH^- remains in the liquid phase and increases the pH value. The amount of rain between tsunami inundation and soil sampling was 70–100 mm (Fig. 6.13b).

If the drainage was poor after immersion in seawater, the pH(H_2O) value was kept low owing to high salt concentration. As the pH(KCl) value is determined in 1 mol L^{-1} KCl, which is not very much different from the concentration in seawater, the difference (ΔpH) between the pH(H_2O) and pH(KCl) values tended to be small in the majority of the muddy tsunami deposits due to restricted drainage (Table 6.2). In

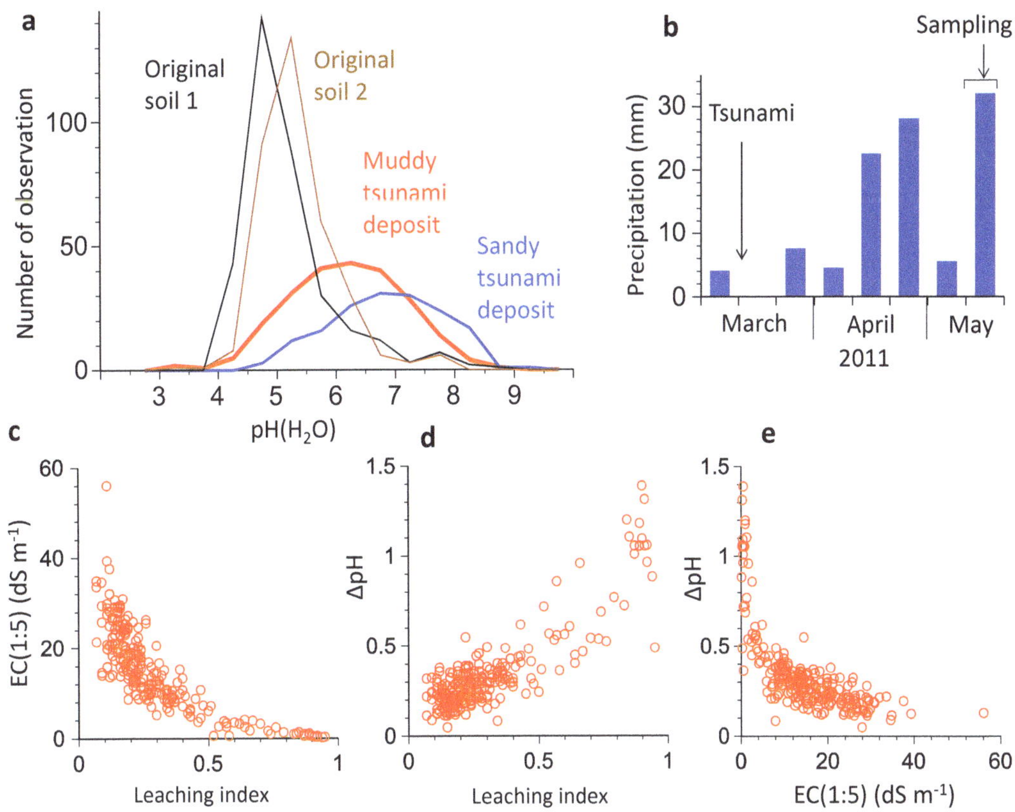

Fig. 6.13 pH(H_2O) values of the tsunami deposits and the original soils. (**a**) Frequency distributions of pH(H_2O) values of the tsunami deposits and the original soils (for example, a plot at pH (H_2O) = 3.25 shows the number of observations ranging within 3 < pH(H_2O) ≦ 3.5), (**b**) precipitation between the day of the tsunami effect and the sampling of the tsunami deposits and original soils, (**c**) relationship between the sum of exchangeable cations (Ca^{2+} + Mg^{2+} + K^+ + Na^+)/(the sum of exchangeable cations (Ca^{2+} + Mg^{2+} + K^+ + Na^+) plus the sum of water-soluble cations (Ca^{2+} + Mg^{2+} + K^+ + Na^+)) (leaching index) and EC(1:5), (**d**) relationship between the leaching index and (pH(H_2O) – pH(KCl)) (ΔpH), (**e**) relationship between EC(1:5) and ΔpH. All of (**c**), (**d**), and (**e**) are for the muddy tsunami deposit (n = 229)

other words, the pH(H_2O) value can be close to the pH(KCl) value for samples if the samples were under restricted drainage.

However, a portion of the muddy tsunami deposit must have been under well-drained conditions. The EC(1:5) value decreased steeply with increase in leaching index (Fig. 6.13c). The intensity of drainage or leaching can be expressed numerically by the leaching index, which is tentatively defined as the sum of exchangeable cations (Ca^{2+} + Mg^{2+} + K^+ + Na^+) divided by the (sum of exchangeable cations (Ca^{2+} + Mg^{2+} + K^+ + Na^+) plus the sum of water-soluble cations (Ca^{2+} + Mg^{2+} + K^+ + Na^+)). The leaching index is based on the fact that water-soluble cations are lost easily with rainwater and drainage, whereas exchangeable cations are lost slowly. The ΔpH value increased with leaching, as shown in Fig. 6.13d. It was also confirmed that the ΔpH of the muddy tsunami deposit is inversely proportional to the EC(1:5) value (Fig. 6.13e).

According to Vorob'eva and Pankova (2008), ESP increased to 50% or higher when the pH (H_2O) value was 9.5 or higher. Similar results were obtained using selected samples from the muddy tsunami deposit. As a large amount of Na_2CO_3 was necessary to raise the ESP values higher than 50%, it was estimated that the ESP values of the tsunami-affected deposit in the humid region hardly rise higher than 50%.

6.2.7 *Desalinization and Restoration of the Tsunami-Affected Farmland*

The tsunami-affected area is under a humid and temperate climate. Natural rainwater will gradually remove salts from the soil. On the other hand, lowland soils, peat soils, and muck soils are distributed in the region (Fig. 6.5c), and the groundwater level is mostly high. Drainage water was pumped out near the major river mouths of the central and southern part of Miyagi Prefecture, although the pumping system was destroyed by the tsunami. Under these conditions, Figs. 6.14 and 6.15 show an example of the vertical distribution of salts in paddy field with the passage of time. This study site is located 3 km inland from the shoreline. The EC(1:5) value of the muddy tsunami deposit (4 cm thick) was 9.2 dS m^{-1} on June 15, 2011 (Fig. 6.14a), and a portion of salts reached a depth of 36 cm from the surface. This date was 3 months after the tsunami, and a total of 250 mm of rainwater had fallen. The EC (1:5) values of the tsunami deposit and the Ap horizon soil were lowered to around 1.5–2 dS m^{-1} by September 4, 2011 (Fig. 6.14b), by natural rainwater (Fig. 6.14d). Although the tsunami deposit was removed from the paddy field after this time, the EC(1:5) values were reduced further to values lower than 0.6 dS m^{-1}, a safe level for mesophytes, by natural rainwater to the depth of 20 cm by June 24, 2012.

After removal of the soluble salts in the liquid phase by natural rainwater, Na^+ held on the exchange sites of the soil still tended to remain (Fig. 6.15b, c) whereas EC(1:5) values were lowered (Fig. 6.14c). The low exchangeable Na^+ values at the depth of around 30–40 cm are due to the insertion of a sand layer (Fig. 6.16a).

To restore the farmland damaged by the tsunami, debris from damaged homes, fallen trees, and excessive salts had to be removed. The tsunami deposits have also been removed from the farmland in Miyagi Prefecture. A portion of the Ap horizon may have been removed during this operation. Irrigation water was finally used to remove salts to a safe level. Soil from nearby mountains was dressed to refill the original farmland level. A reduction in soil fertility was improved by the application of fertilizer.

There have been many studies pertaining to salt-affected farmlands and rehabilitation (Agus and Tinning 2008; Nakaya et al. 2010). The drainage system, including the underground one, plays an important role in desalinization by both rainwater and irrigation water. In addition, in this study region, the ground along the coastal areas

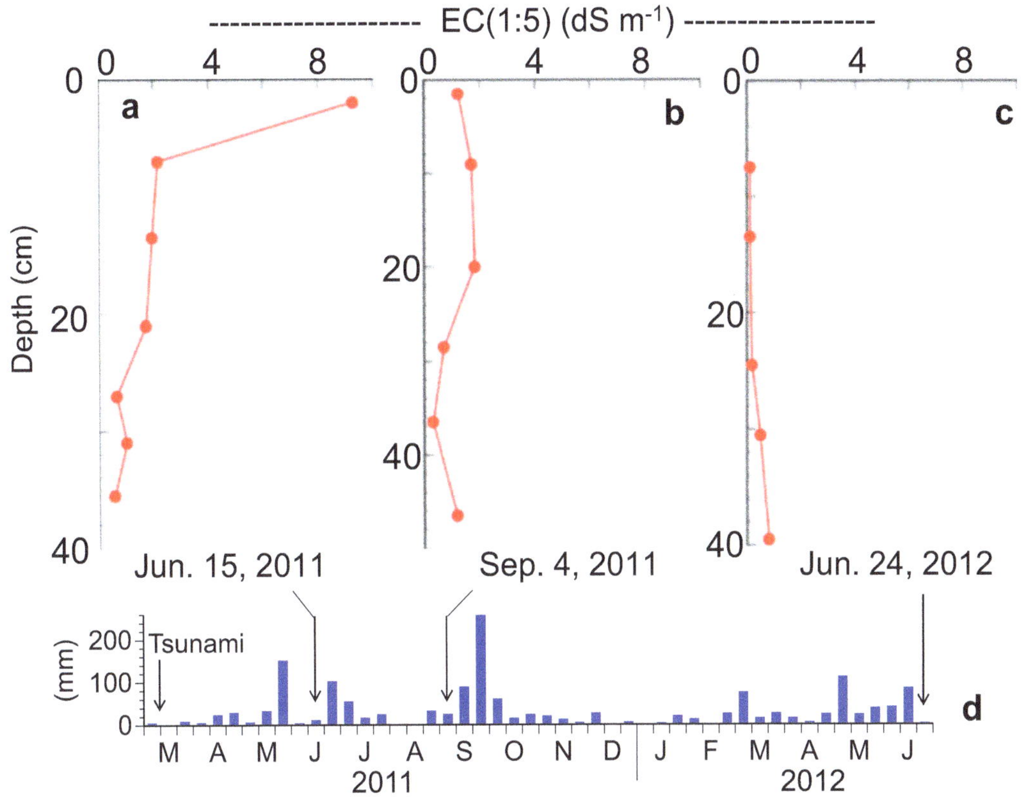

Fig. 6.14 Changes in the vertical distribution of EC(1:5) in the tsunami-affected paddy field soils under only natural rainwater. The uppermost tsunami deposit was removed artificially between September 4, 2011, and June 24, 2012. These three small pedons were sampled from two neighboring paddy fields. (**a**) June 15, 2011, (**b**) September 4, 2011, (**c**) June 24, 2012, (**d**) precipitation between March 2011 and June 2012. For an example of the soil profile of (**a**, **b**, and **c**), see Fig. 6.16a

sank by several tens of centimeters due to the earthquake. In these areas, dikes, drainage, and the pumping system should be restored.

Even when salt removal was successful in the upper part of the farmland soil, soluble salts partly remained in the lower part of the soil (Fig. 6.14c). In the case of the paddy fields, if irrigation water is supplied, the detrimental effect of the salt remaining in the subsoil may be negligible for rice growth because the salt may move further downward due to the percolation of the irrigation water. However, the salt remaining in the lower part of the soil may affect the growth of soybean, which is often cultivated in the drained rice fields due to rice production control. The salt tolerance of plants varies widely. During the survey in May 2011, barley had spikes in the tsunami-affected fields, although the plant height was short because spring fertilizer had not been applied.

By the spring of 2012, 39% of the damaged farmland had been restored (Ministry of Agriculture, Forestry and Fisheries, Japan 2012). In Fukushima Prefecture, restoration has been delayed due to contamination by radioactive materials from the nuclear power plant accident.

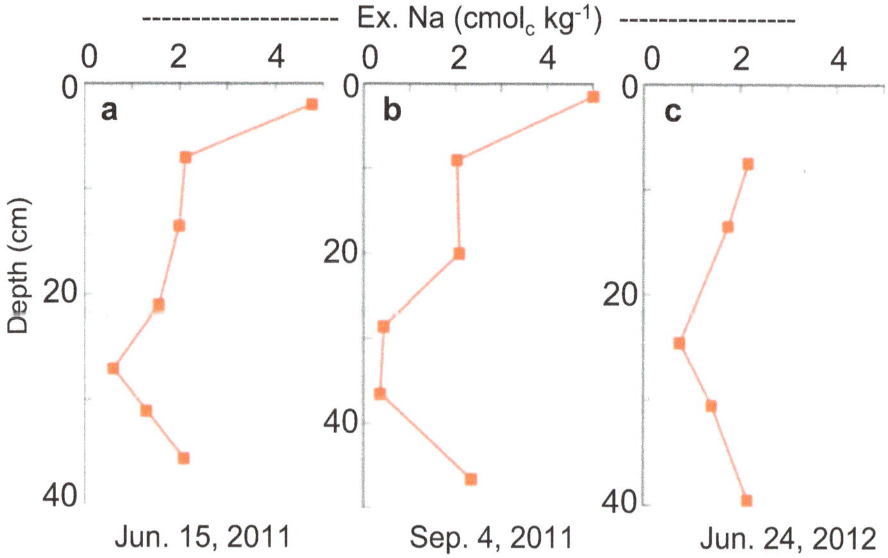

Fig. 6.15 Changes in the vertical distribution of exchangeable Na^+ with time in the same three pedons shown in Fig. 6.14. The uppermost tsunami deposit was removed artificially between September 4, 2011, and June 24, 2012

Fig. 6.16 Geological records of huge tsunamis. (**a**) Soil profile inundated by the 2011 tsunami, (**b**, **c**, and **d**) sand fractions of tsunami deposit (2011), Towada-a tephra (AD915), and tsunami deposit (AD869), respectively

Huge tsunamis have been recorded in many places in the world (Okal 2011; Satake et al. 2011). The Indian Ocean tsunami of 2004 was a recent huge tsunami, such huge tsunamis have occurred repeatedly in the region in the past (Monecke et al. 2008). The evidence of previous tsunamis is recorded as relatively well-sorted sandy layers.

According to geological records (Minoura et al. 2001), the present study area has also been struck repeatedly by huge tsunamis. Figure 6.16a shows a soil profile including an old sandy tsunami deposit. Although the present tsunami deposits are removed in Fig. 6.16a, the sand fraction of the 2011 tsunami deposit is shown in Fig. 6.16b. At the depth of 20–30 cm from the surface, the Towada-a (To-a) tephra, an indicator of the soil age of AD 915, occurs as light-brown patches (Fig. 6.16a). The underlying sand layer is the tsunami deposit of AD 869 (Minoura et al. 2001). The sand fraction particles shown in Fig. 6.16b, d have similar subrounded shapes, which suggests sea sand. The ^{14}C age of plant fragments collected from the soil horizon beneath the tsunami deposit layer (AD 869) is consistently old (1450 BP) (Kanno et al. 2013).

6.3 Radiocesium

Radiocesium is a radioactive element that is an artificial contaminant of the soil–biota system. Although radiocesium is sorbed strongly by soil inorganic constituents, especially weathered mica (Fig. 2.21), a trace fraction of the radiocesium is absorbed by plants, fungi, and other organisms. The half-lives of ^{134}Cs and ^{137}Cs are 2.1 and 30.2 years, respectively.

Radiocesium was introduced into soils all over the world in the 1950s–60s due to atmospheric nuclear testing and it remains detectable in soil. Furthermore, radiocesium was released into the environment by the accidents at the nuclear power plants in Chernobyl (1986) and Fukushima (2011).

Although Steinhauser et al. (2014) estimated that the total release of radionuclides from the Fukushima Daiichi Nuclear Power Plant (FNPP) accident (approximately 520 (340–800) PBq) was approximately one order of magnitude lower than the release from Chernobyl (approximately 5300 PBq), both accidents have been rated as “major accidents” by the International Atomic Energy Agency. This section describes the radiocesium contamination of soil from the FNPP accident, which was caused by the huge tsunami that occurred on March 11, 2011.

6.3.1 *Horizontal Distribution of Radiocesium*

Highly contaminated areas are distributed in an inverse L-shaped pattern in the northwestern direction from the FNPP (Fig. 6.17), although more than 80% of the entire radioactive elements was released to the Pacific Ocean. Factors related to the

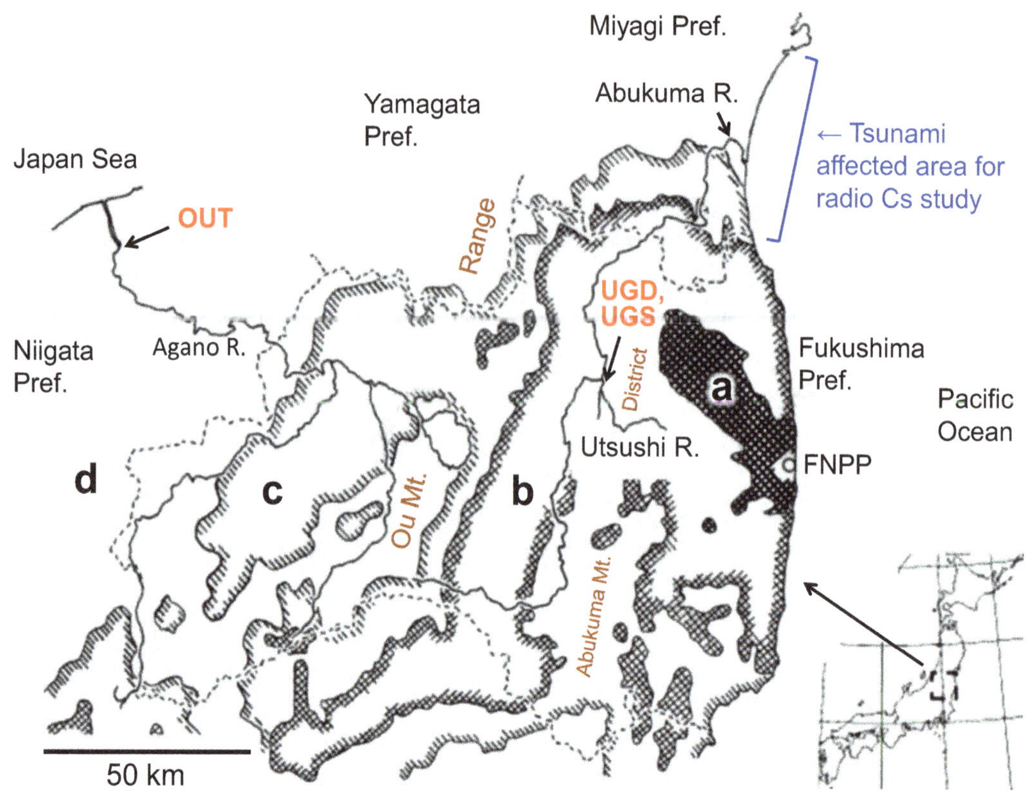

Fig. 6.17 Tracing of deposition densities of $^{134}Cs + ^{137}Cs$ (Bq m^{-2}, a > 1000 k; 1000 k > b > 60 k; 60 k > c > 10 k; d ≦ 10 k) made by using airborne monitoring results (converted as of October 13, 2011) of the Geospacial Information Authority of Japan (2014). **UGD** and **UGS**: sampling sites in the Utsushi River of Fukushima Prefecture, **OUT**: sampling site in the Agano River of Niigata Prefecture. Branches of Abukuma River and Agano River are not shown for simplicity, although they are numerous. **FNPP**: Fukushima Daiichi Nuclear Power Plant of the Tokyo Electric Power Co., Inc. The study site of radiocesium in the tsunami-affected area is shown in blue

radiocesium distribution are the wind direction to the northwest of the FNPP and the topography of the area (Hirose 2016; Saito et al. 2015). On March 15, 2011, a large amount of radioactive elements was released from the FNPP. The wind direction at that time was from the coastal area to the Abukuma Mountain District. When the wind reached the Ou Mountain Range, the wind was directed to a southern direction. As the weather in these areas was rainy or snowy on the day of the accident, radiocesium was deposited by wet scavenging (Saito et al. 2015).

6.3.2 Vertical Distribution of Radiocesium in Soil

Fixation of radiocesium is so strong that most of the radiocesium deposited on the surface of soil hardly moves and remains near the soil surface. The vertical distribution of radiocesium was determined in the tsunami-affected areas in Miyagi

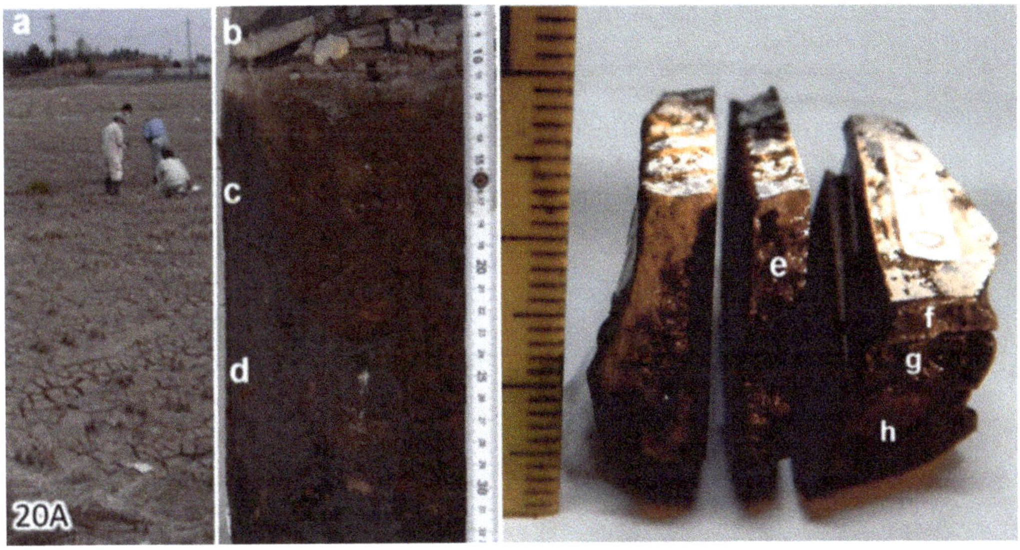

Fig. 6.18 An example sampling site for the radiocesium study. (**a**) Tsunami-affected farmland (sampling site No. 20A), (**b**) muddy tsunami deposit, (**c**) original soil 1, and (**d**) original soil 2 of this sampling site, (**e**) vertical section, (**f**, **g**, and **h**) horizontal sections (2–3 mm thick) of the muddy tsunami deposit (**b**)

Prefecture. The area is located between 50 and 100 km north of the FNPP (Fig. 6.17), and it is the same as in Fig. 6.4b. Although the tsunami deposits were mostly removed from the farmland in Miyagi Prefecture, the distribution of radiocesium in these soils may need to be taken into account in subsequent management of the soils. Figure 6.18a shows one of the sampling sites. The method used to collect samples of the tsunami deposits and the original soils was the same as that described in Sect. 6.2.1. Figure 6.18b, c, d show muddy tsunami deposit, original soil 1, and original soil 2, respectively. The muddy tsunami deposit (Fig. 6.18b) was cracked by desiccation at the time of sampling. The sandy tsunami deposit at this site was too thin to sample separately.

Fine-earth fractions from a total of 17 fields, seven from Sendai, five from Natori, and five from Watari-Yamamoto (open red circles in Fig. 6.19a), were packed into commercial U-8 plastic containers to the height of 12 mm, and radiocesium activity concentrations were determined using gamma-ray spectrometry (see Sect. 1.4) (Nanzyo et al. 2015).

The muddy tsunami deposit had the highest radiocesium activity concentration at all 17 sites (Fig. 6.18b, c, d). Activity concentrations of radiocesium lower than those of the muddy tsunami deposit were also detected in the sandy layer at several sites. The radiocesium activity concentrations in the underlying original soils 1 and 2 were very low. These results show that the muddy tsunami deposits fixed most of the radiocesium deposited in this area. Thus, the distribution of radiocesium in the muddy tsunami deposit was examined further.

Soil sections were prepared by cutting vertical layers from a piece of cracked muddy tsunami deposit after hardening the piece with resin (Fig. 6.18e). The

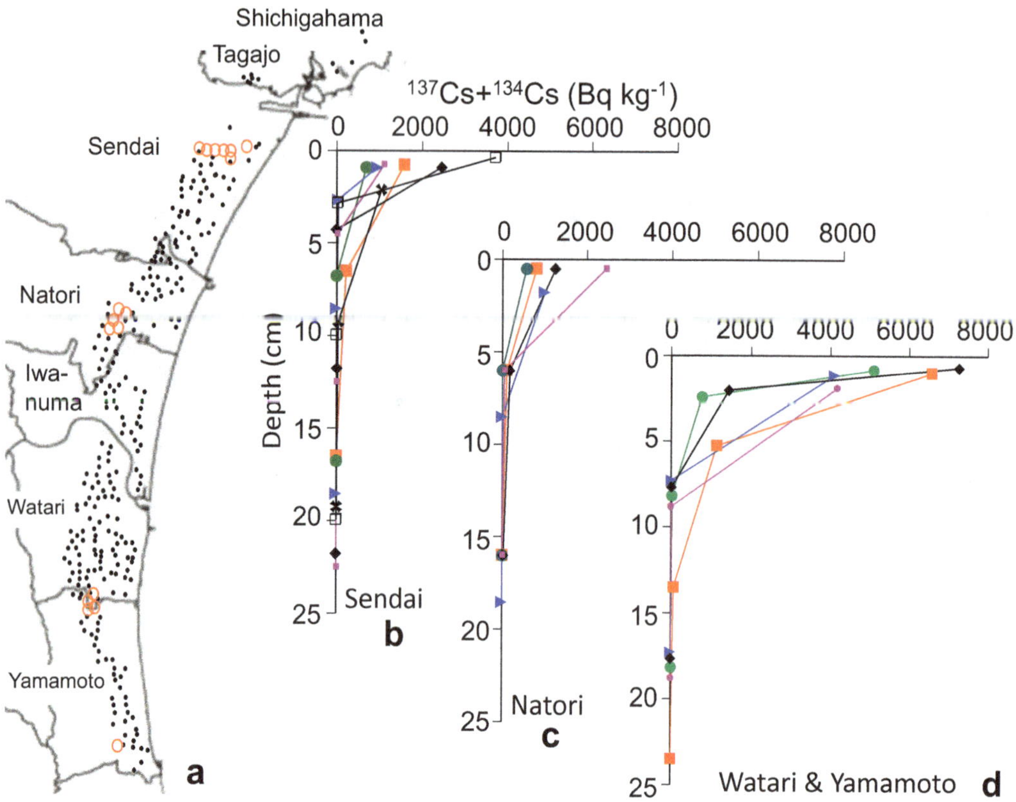

Fig. 6.19 Vertical distribution of radiocesium. (**a**) Map showing the 17 locations of tsunami-affected soils used for the radio Cs study (red open circles), (**b**, **c**, and **d**) vertical distribution of radio Cs concentration in the Sendai, Natori, and Watari & Yamamoto areas, respectively. The uppermost plot at each location show the radiocesium concentrations of the muddy tsunami deposit. The deepest and the second deepest plot at each location show the radiocesium concentrations of the original soil 2, and 1, respectively

sections were attached to one side of a strip of double-sided adhesive tape, attached to cardboard using the other side of the tape, and then covered by a thin plastic sheet. The cardboard was placed on an imaging plate (IP) for 30 days to detect radioactivity (see Sect. 1.4).

The IP results of the vertical sections revealed two distribution patterns of radioactivity (Fig. 6.20). One pattern was a line of spots with high radioactivity near the surface of the muddy layer. The radioactivity of these spots was sufficiently high to indicate that the radioactivity was from radiocesium. The distribution of these spots suggests that a significant fraction of the radiocesium had been deposited as particulate material (Adachi et al. 2013) and that deposition of the radioactive particulate material occurred after the sedimentation of the muddy tsunami deposit. The other pattern was a very faint, layer distribution at the upper part of the vertical section. If this radioactivity was also from radiocesium, the radiocesium may have been deposited in a soluble form and become fixed by suspended soil materials in the inundation water during sedimentation (Mukai et al. 2016). It was not clear whether a few spots showing radioactivity at the side edge of "2A" from Watari &

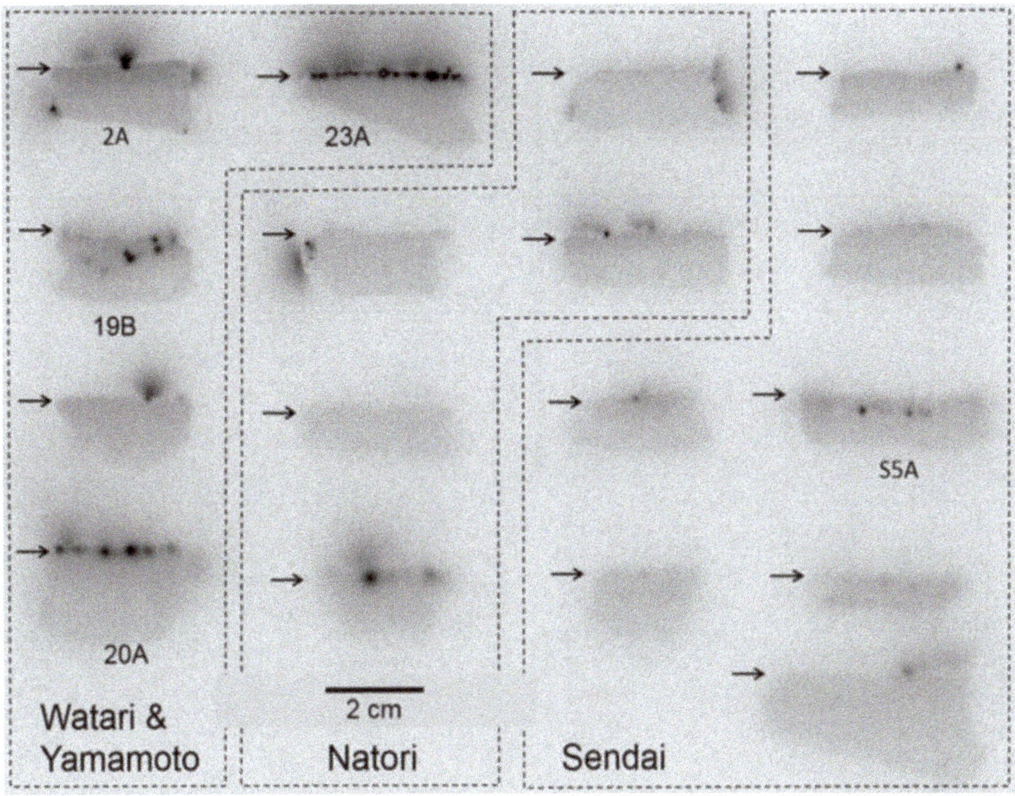

Fig. 6.20 Distribution of radioactivity determined by an imaging plate using vertical sections of muddy tsunami deposits from the Sendai, Natori, and Watari & Yamamoto areas. See Fig. 6.18e for information about the samples used. Arrows show the positions of the surface of the muddy tsunami deposits

Yamamoto, for example, were a type of artifact formed during handling of the sample or were from some other cause.

Because the radioactivity of spots of samples 20A and 23A was so intensive, the vertical distribution of radioactivity was examined using 2–3 mm thick horizontal sections of four samples (2A, 23A, 20A, and S5A), as exemplified in Fig. 6.18f, g, h. The same gamma-ray spectrometry used for Fig. 6.19 was used for radiocesium measurement of the horizontal sections. The results are shown in Fig. 6.21. As shown by the many radioactive spots in "20A" and "23A" in Fig. 6.20, higher radioactivity concentration was obtained from the surface sections (Fig. 6.21), suggesting high radioactivity of the spots near the surface of the sections. In the cases of "2A" and "S5A", as there were fewer radioactive spots, the radioactivity concentrations of the surface sections were not as distinctively high as those of "20A" and "23A". The radiocesium concentration of the second layer of "2A" appears higher than that of the third section, suggesting that the faint layer distribution of radioactivity of "2A" in Fig. 6.20 is due to radiocesium.

Using "19B" and "20A" of the muddy tsunami deposits shown in Fig. 6.20, six particle size fractions were prepared after H_2O_2 digestion followed by wet sieving and a sedimentation method. The radiocesium concentrations of the six particle size

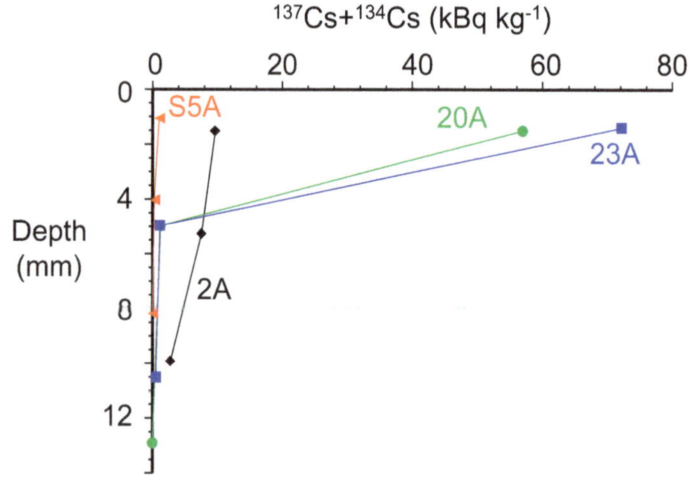

Fig. 6.21 Vertical distribution of radiocesium concentration in the muddy tsunami deposit. For the sample preparation, see Fig. 6.18f, g, h

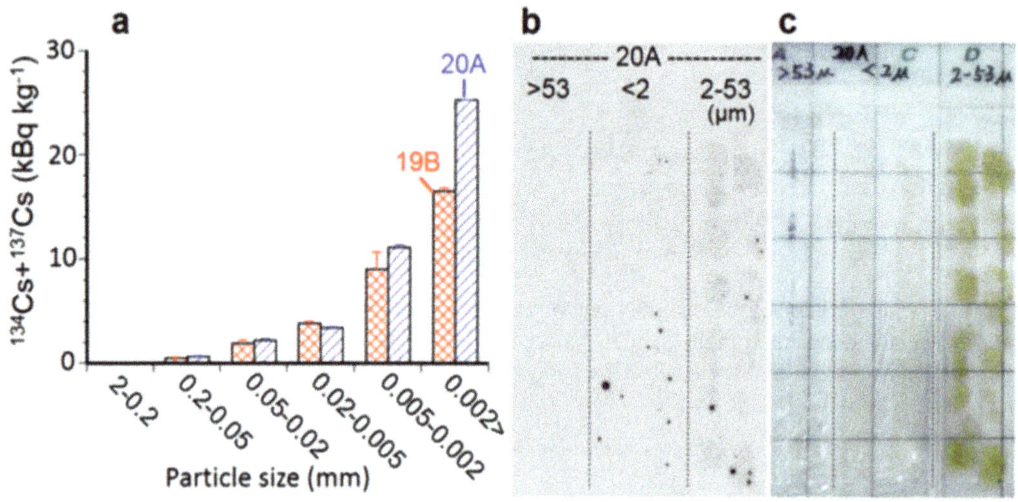

Fig. 6.22 Radiocesium in the muddy tsunami deposits. (**a**) Changes in radiocesium concentrations with particle size, (**b**) radioactivity of three particle size fractions using an imaging plate (IP), (**c**) optical photograph of the three particle size fractions used for radioactivity detection using IP. Samples 19B and 20A are the same as those in Fig. 6.20

fractions obtained using the same method as in Fig. 6.19 are shown in Fig. 6.22a. Among the six particle size fractions, the radiocesium concentration was highest in the clay fraction (< 2 μm), and it decreased with increasing particle size. The radioactivity of the three particle size fractions shown in Fig. 6.22c was also detected using the same IP method used for Fig. 6.20. In the two particle size fractions of <2 μm and 2–53 μm, high radioactivity particles were significantly detected, whereas few were detected in the >53 μm particle size fraction. Hence, the highly radioactive particles in "20A" are smaller than 53 μm in diameter. The highly radioactive particles appeared to be stable during H_2O_2 treatment and particle size fractionation, but they disappeared when the <2 μm fraction was treated with hot 6 mol L^{-1} KOH.

Radioactivity was also detected in the particle size fractions larger than 53 μm (Fig. 6.22a). Weathered biotite grains larger than 53 μm may be included in the muddy tsunami deposit because they can fix radiocesium.

6.3.3 Fixation of Cesium Ion by Soil

The mechanism of radiocesium fixation by weathered biotite is similar to that for potassium, which is shown in Fig. 3.11. Although a large amount of cesium was added, the fixed cesium can be detected by EDX. Figure 6.23a shows examples of partially weathered biotite particles. The particles still include potassium. After washing the weathered biotite particles twice with 2 mol L^{-1} CsCl solution, followed by washing with water, and air-drying, X-ray counts from Cs(Lα) are detectable in the EDX spectrum (Fig. 6.23b). The Cs(Lα) peak is not detected in the EDX spectrum of the untreated sample (Fig. 6.23c). The fixed Cs is also detectable in a Cs element map (Fig. 6.23d). Cs is distributed over the whole area of the weathered biotite particles, where K is also detected (Fig. 6.23e). Under the accelerating voltage of 15 kV, X-rays from the depth of about 1 μm, which is much thicker than the basal spacing (1 nm) of biotite, are detected (see Sect. 1.4).

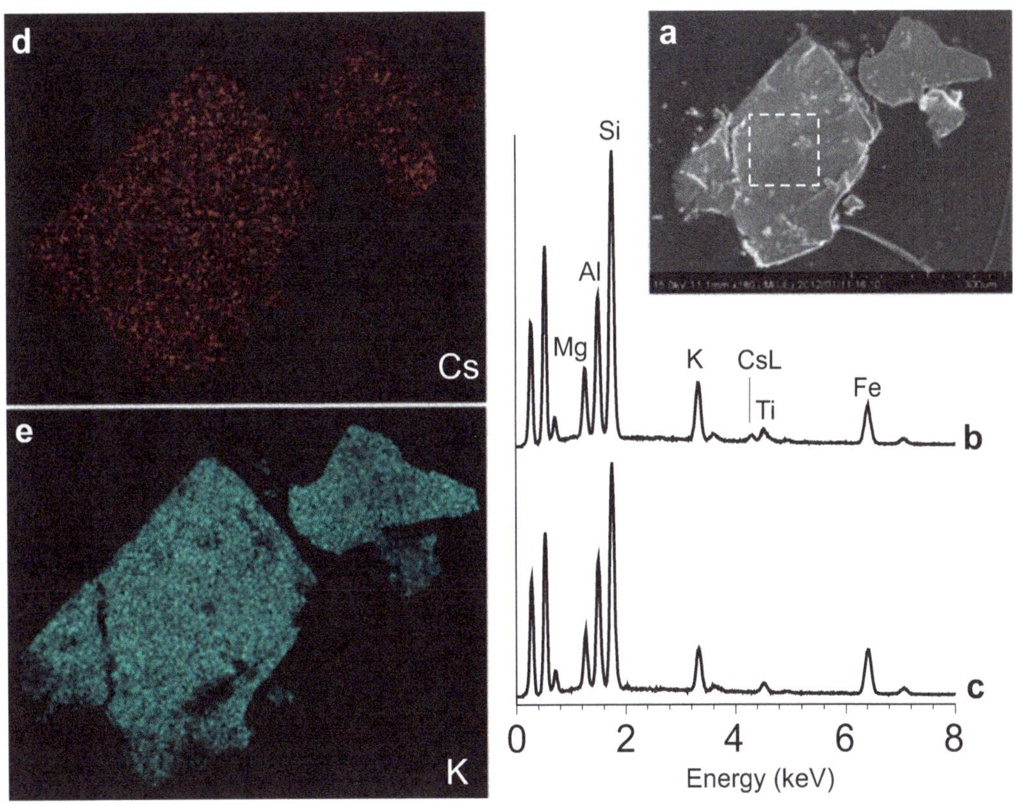

Fig. 6.23 Fixation of cesium ion by weathered biotite. (**a**) SEM image of weathered biotite particles picked up from a side bar deposit (2nd layer, Fig. 6.25a) of the Utsushi River, (**b** and **c**) EDX spectra of Cs-sorbed and untreated weathered biotite, respectively, (**d** and **e**) element maps of Cs and K, respectively, of the Cs sorbed weathered biotite particles. The EDX spectrum (**b**) was obtained from the dashed square in (**a**)

6.3.4 Transportation of Radiocesium in Rivers Estimated from Side Bar Deposits

The radiocesium deposited on a soil surface is fixed by soil particles (Shiozawa 2013; Nakao et al. 2014). However, when the soil particles move because of erosion, the radiocesium moves with the particles. High radiocesium concentration was detected in the water treatment residue (WTR) (Ippolito et al. 2011) from the Aganogawa water purification plant which locate near Ounbashi sampling site (OUT) in Fig. 6.17, 170 km away from FNPP. In the case of rivers, soil particles move with the water in suspended form, particularly during heavy rains (Evrard et al. 2014), and are eventually deposited as bars. Bars are often found in slow-flowing, shallow parts of a river.

The Fukushima nuclear accident in 2011 was followed by snowmelt, heavy rain, and a typhoon, which caused rising levels of muddy water in rivers and the formation of bars. Information about the transportation and deposition of soil particles with radiocesium may be obtained from these side bars.

Contrasting vertical distributions of radiocesium from soil can be found in the river side bar deposits. On December 24, 2011, there were side bar deposits with a depth of 1–2 m on the right side of the Utsushi-Gawa dam (UGD) (Fig. 6.24a). Soil samples collected from the dam lakeshore (Fig. 6.24b) (UGS), which is located on the right side of the UGD, showed an ordinary vertical distribution pattern of radiocesium: the surface soil had high radiocesium concentration and the radiocesium concentration decreased steeply with increase of depth (Fig. 6.24c).

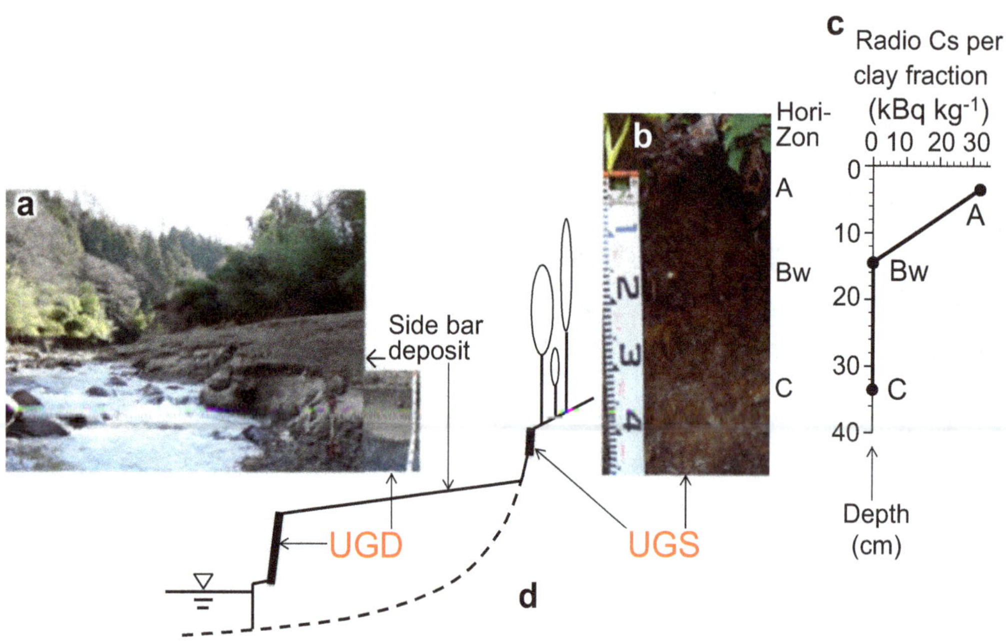

Fig. 6.24 Contrasting vertical distribution of radiocesium between soil and side bar deposit. (**a**) Landscape and profile of side bar deposit at the Utsushi-Gawa dam (UGD), (**b**) soil profile of the Utsushi-Gawa dam lakeshore (UGS), (**c**) vertical distribution of radiocesium concentration per clay fraction of UGS, (**d**) schematic diagram showing the relationship between UGD and UGS

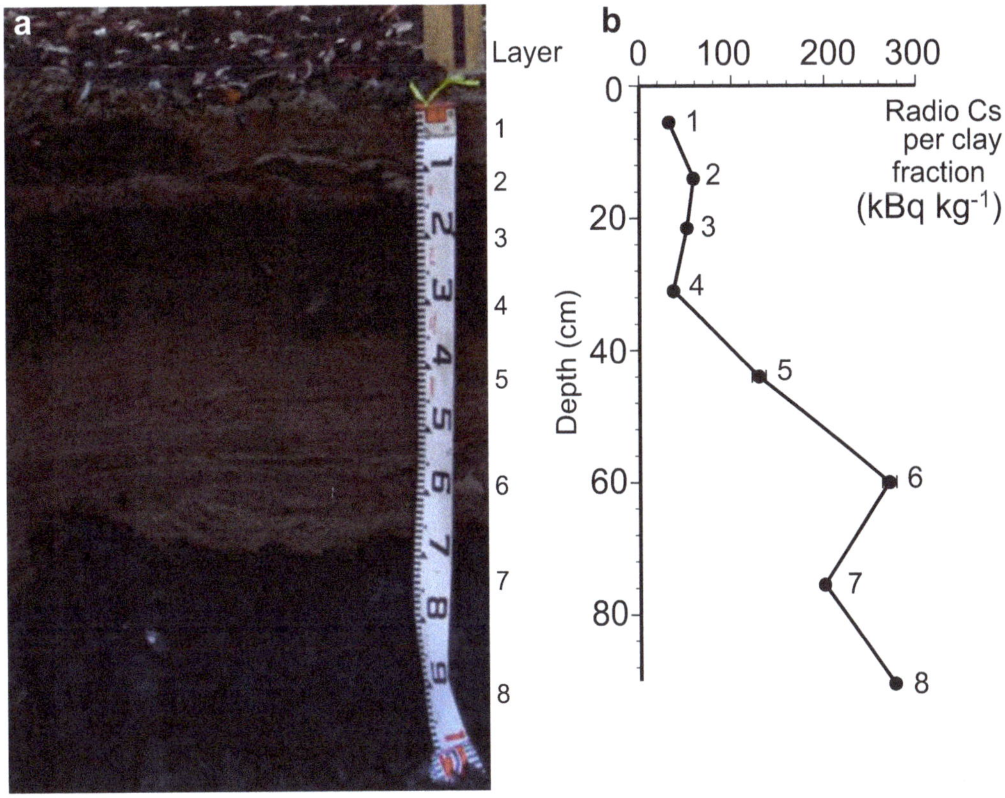

Fig. 6.25 Radiocesium in the side bar deposit. (**a**) Profile of the side bar deposit of the Utsushi-Gawa Dam (UGD), (**b**) vertical distribution of radiocesium concentration per clay fraction of the side bar deposit

High concentrations of radiocesium were detected especially from the 8th layer of the UGD site (UGD-8) (Fig. 6.25b) as the Utsushi River flows from the Abukuma Mountain District close to the FNPP (Fig. 6.17). The radiocesium concentration differed widely among layers. UGD-5 and UGD-6 were artificially subdivided layers with a similar appearance in the field, yet with a fine cumulative stratification (Fig. 6.25a). Layers UGD-7 and UGD-8 were similarly separated. The vertical distribution of the clay content also differs widely. High clay content appears to coincide with high radiocesium concentration, suggesting that clay minerals are the most important binder of radiocesium. In addition, the radiocesium concentration decreased upward, with lower radiocesium concentrations in UGD-1 than in UGD-7, despite almost identical clay content (Nanzyo et al. 2014).

The side bar deposits (UGD of Fig. 6.25 and OUT of Fig. 6.26) were thick and included layers with various clay content. Although some sand-sized vermiculite can carry radiocesium, clay minerals that receive fallout of radiocesium play a major role in carrying radiocesium (Tsukada et al. 2008). However, the sand content of the side bar deposit is changeable depending on the velocity of water flow, and the clay content may be diluted with sand. The concentration of radiocesium per clay fraction should more directly reflect the effect of fallout, especially in the sandy layers of the side bar deposits. Furthermore, the radiocesium concentration per clay fraction can

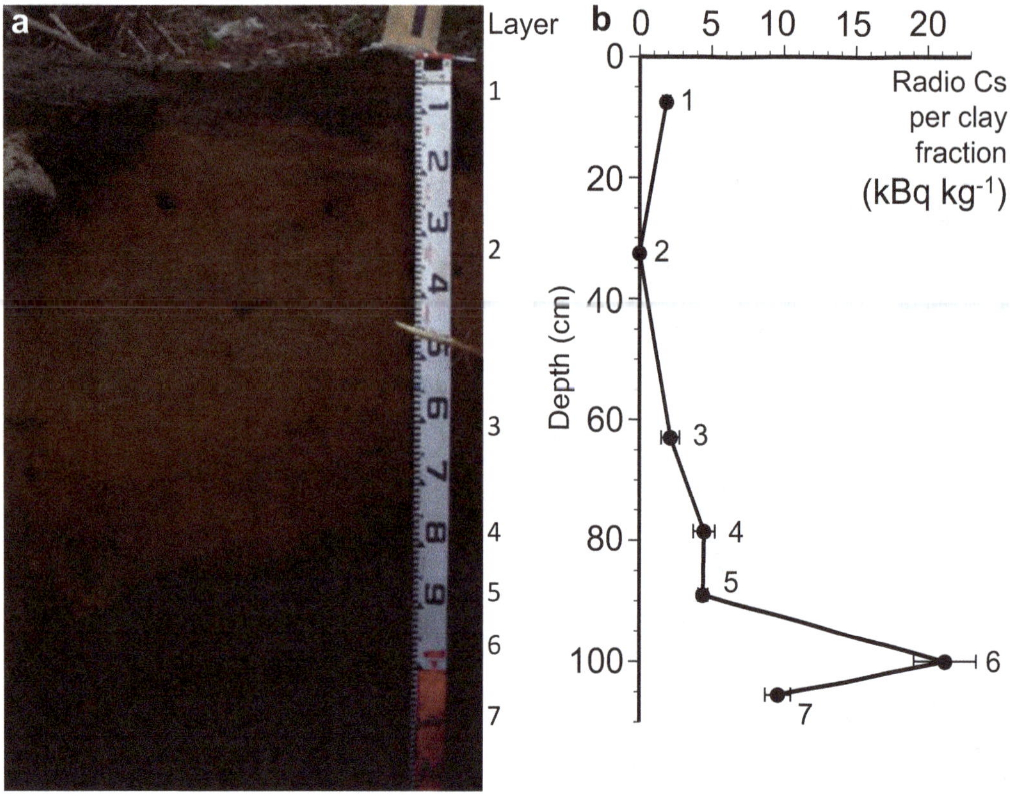

Fig. 6.26 Radiocesium in side bar deposit. (**a**) Profile of the side bar deposit of the Ounbashi Aganogawa (OUT), (**b**) vertical distribution of radiocesium concentration per clay fraction of the side bar deposit

be more comparable to that of the WTR from a water purification plant, considering their particle size. Hence, the concentration of radiocesium per clay fraction was calculated by dividing the concentration of radiocesium in each layer by the clay content of the layer. The results are shown in Figs. 6.25b and 6.26b. The radiocesium per clay fraction was more than five times greater than that in the fine earth fraction (particle diameter ≦ 2 mm). For UGD, the concentration of radiocesium in the clay fraction was high in the earlier (lower) muddy layers, UGD-7 and UGD-8. High concentrations of radiocesium per clay fraction continued into the sandy UGD-6 layer, after which the concentrations decreased (Fig. 6.25b).

Similar results were obtained for OUT, as shown in Fig. 6.26; however, the maximum concentration of radiocesium per clay fraction (OUT-6, ca. 20 kBq kg^{-1}) was an order of magnitude smaller than that of UGD.

The maximum concentration of radiocesium per clay fraction of OUT (Fig. 6.26b) was similar to that of the WTR from the Aganogawa water purification plant, 35,400 Bq kg^{-1}, sampled during August 24–30, 2011 (Yomiuri Newspaper 2011). This WTR was generated by the water purification treatment during the period of December 2010–May 2011. If the particle size of the WTR generated from the water purification plant is comparable to the size of the clay fraction, it would justify the similarity of these values. The radiocesium per clay fraction of OUT decreased in the shallow layers that were deposited later. Similarly, the

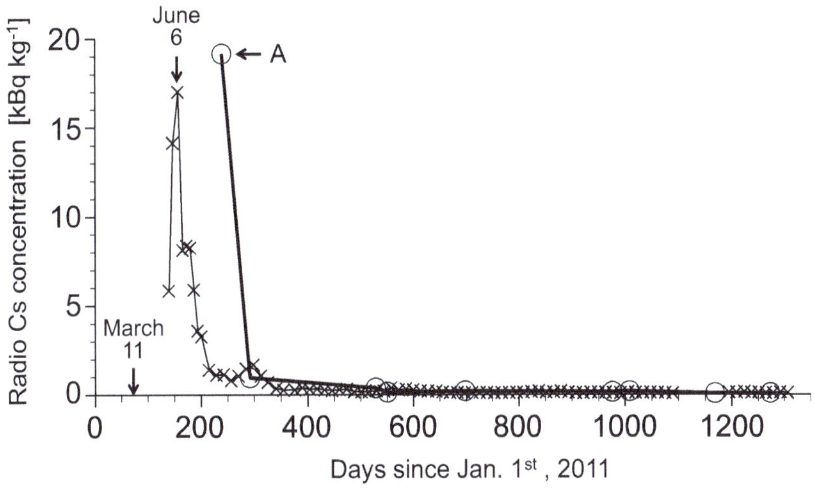

Fig. 6.27 Changes in radiocesium concentration of water treatment residue from the Aganogawa (○) and Manganji (×) water purification plants with days since January 1, 2011 (DSJF). The plots shown by (○) are midpoints calculated from maximum and minimum values. For instance, plot A is the midpoint between 2890 and 35,400 Bq kg^{-1} observed during 236 and 242 DSJF. The DSJF of the water intake from the Agano River was 45–60 days and 120–240 days earlier than the measurement of radiocesium at the Manganji and Aganogawa water purification plants, respectively

concentration of radiocesium in the WTR decreased steeply, although values of hundreds of Bq kg^{-1} lasted for several months, according to the Aganogawa (Niigata City 2014) and Manganji water purification plant report (Niigata City 2014b), as shown in Fig. 6.27. Many data are available from the Manganji water purification plant where WTR is produced by machine dewatering every 1.5–2 months. The frequency of WTR production is less at the Aganogawa water purification plant where WTR is produced by sun-drying every several months. Results similar to those of Fig. 6.27 were reported from water purification plants in Saitama Prefecture (Saitama Prefecture 2014).

Radiocesium may have deposited on riverbeds with little water or alternatively on the snow of riverbeds and riverside areas. Hence, the high radiocesium concentration in WTR and in the lower layers of the relatively new side bar deposits at UGD and OUT may be related to snowmelt-related transport of soil particles or water containing radiocesium (Nanzyo et al. 2014).

Further readings are Nakanishi and Tanoi (2013, 2016).

6.4 Phosphates Related to Soil-Plant Systems

The phosphorus concentration of Earth's crust is estimated to be 0.1%. Phosphorus is one of the major essential elements of the organisms, and it is one of the elements that limit the amount of biota. In farmlands, phosphate (P) fertilizer is effective for increasing crop production. However, as soils more or less fix P, the efficiency of P fertilizer is generally lower than those of nitrogen and potassium fertilizers. Thus, increasing the efficiency of P fertilizers, typically lower than 20%, is a concern. P accumulates in the plow layer soils of farmlands, and with an increase in P accumulation, a portion of the P is released gradually to rivers, lakes, and bays causing eutrophication. The natural resource of P fertilizers is phosphate rock. Economically mineable phosphate rock appears to be limited (Amundson et al. 2015), although estimates of the world P reserves require further research (Edixhoven et al. 2014). As 70–80% of phosphate rock is consumed as fertilizers, a sustainable management method of P or recycling of P is desirable for farmlands. This section introduces the forms of P related to soil–plant systems, except for those discussed in Chap. 5.

6.4.1 Apatite and Related Reactions

Apatite is a major mineral of phosphate rock. The general chemical formula of apatite is $Ca_5(PO_4)_3X$ ($X = F^-$, Cl^-, OH^-, and others). Apatite is largely divided into apatite of sedimentary and igneous origins. Apatite of sedimentary origin is microcrystalline, whereas apatite of igneous origin is highly crystalline.

Figure 6.28 shows an example of a sedimentary-origin apatite from Florida, USA. As shown in the optical micrograph (Fig. 6.28a), the major particles are subrounded and have various sizes, although some are broken. The particles have mostly grayish color, but some are light-brown or some other color. These particles are apatite, except for a small amount of colorless transparent quartz. As shown in the magnified SEM image (Fig. 6.28b), the sedimentary apatite appears to be an aggregate of microcrystals. According to the EDX spectrum (Fig. 6.28c), the Florida apatite has a significant amount of F^-. The powder XRD pattern (Fig. 6.28d) is identical to the pattern of the reference fluoroapatite (Fig. 6.28e), except that a small amount quartz is included.

Apatite is converted to fine powder or to more soluble P fertilizers, such as $Ca(H_2PO_4)_2{\cdot}H_2O$, $(NH_4)_2HPO_4$, and others, by adding acids.

Fresh apatite particles are found in young volcanic ash soils (Fig. 6.29) (Nanzyo et al. 1997; Nanzyo and Yamasaki 1998; Nanzyo et al. 2003) or granitic soils (Fig. 2.18). These have an igneous origin and are more crystalline than those of sedimentary origin. As apatite is a minor constituent in volcanic ash and has high particle density, heavy liquid is used to concentrate apatite particles. An example of volcanic ash, from Mt. Pinatubo (1991), is shown in Fig. 6.29 (Nanzyo et al. 1997). An SEM image of the heavy fraction of the Pinatubo ash is shown in Fig. 6.29a, and four P-rich particles were identified using element maps. The particle labeled “b” and one other particle are almost entirely composed of apatite because their shapes in the element maps of P (Fig. 6.29d) and Ca (Fig. 6.29e) are very close to those in the SEM image (Fig. 6.29a). The particle labeled “c” is a composite particle with a Fe–Ti oxide (Fig. 2.15), as shown in Fig. 6.30. The EDX spectra (Fig. 6.29b, c) of the dashed squares labeled “b” and “c” in Fig. 6.29a support the conclusion that these grains are apatite.

Fig. 6.28 Apatite from Florida, USA. (**a**) Optical micrograph, (**b**) SEM image, (**c**) EDX spectrum, (**d**) powder XRD pattern, (**e**) reference powder XRD pattern. (Lehr et al. 1967)

Acid solutions are used for the evaluation of plant-available P in soil. For example, in the Truog method, a solution containing 1 mmol L^{-1} H_2SO_4 and 3 g L^{-1} of ammonium sulfate is used to extract P from soil. Although apatite is not highly available for many crop plants, apatite is soluble in the Truog extracting solution (Fig. 6.30). Figure 6.30a shows a magnified composite particle of apatite and a Ti–Fe mineral. After obtaining the SEM image (Fig. 6.30a), the sample was treated with the Truog extracting solution, and a part of apatite apparently dissolved (Fig. 6.30b). The thin transparent film around the dissolved apatite particle in Fig. 6.30b corresponds to vacuum-evaporated carbon. Although the apatite particle was dissolved significantly by this treatment, the chemical composition of the remaining part represented by a dashed square of Fig. 6.30d in Fig. 6.30b was still the same as before the treatment (Fig. 6.29c). The Fe-Ti oxide in contact with the apatite was not affected by this treatment.

The Truog P levels of tephra depend on the rock type (Fig. 6.30e). The fresh rhyolitic to andesitic tephra showed Truog P levels higher than 100 mg P_2O_5 kg^{-1}, whereas the basaltic and basaltic-andesitic tephra showed only low Truog P values. These results are consistent with those obtained by Green and Watson (1982), who reported that apatite or P_2O_5 solubility in silicate melts increases with decreasing SiO_2 content and that crystallization of apatite cannot occur in mafic and low-P

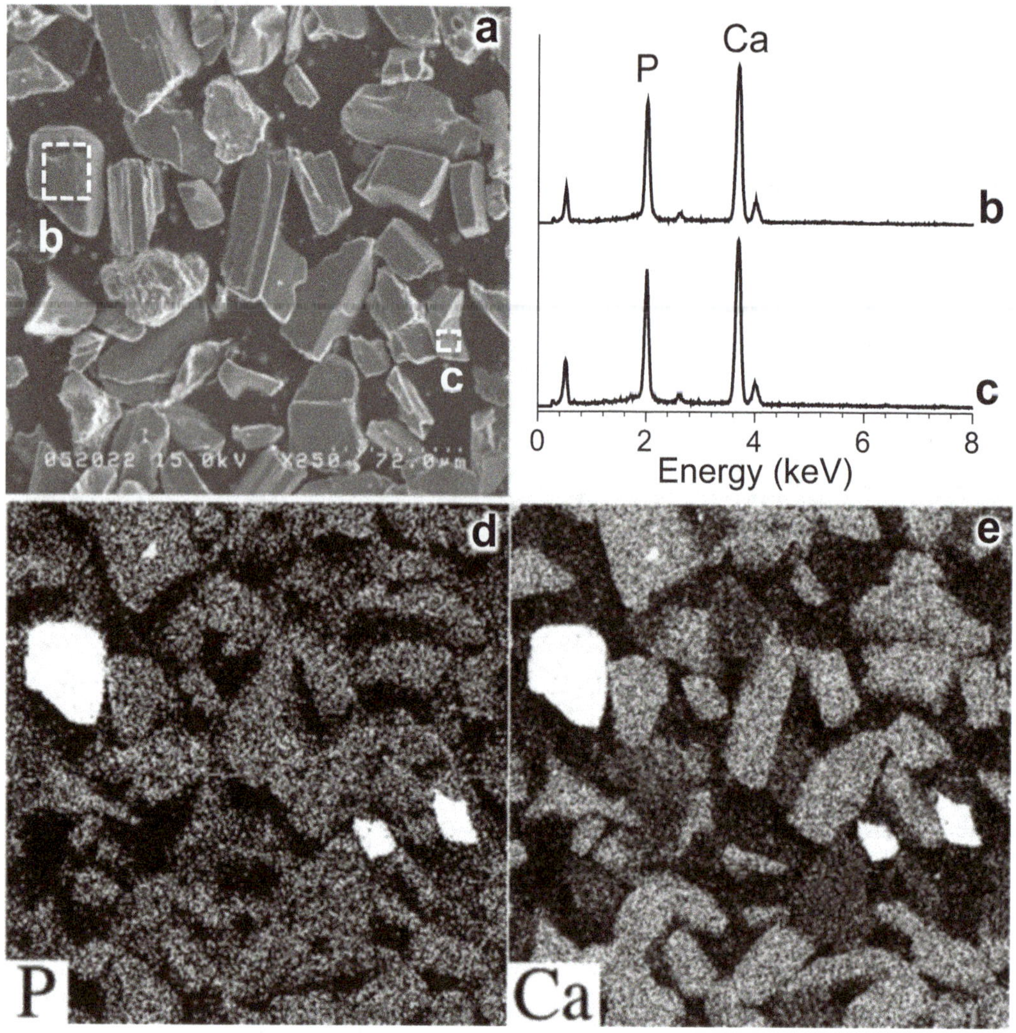

Fig. 6.29 Apatite particles in the Mt. Pinatubo volcanic ash (1991). (**a**) SEM image of the fraction with particle size less than 0.05 mm and specific gravity greater than 2.8 Mg m^{-3}, (**b** and **c**) EDX spectra of the dashed squares (**b**) and (**c**) in (**a**), (**d** and **e**) element maps for P and Ca, respectively

magmas with less than about 2 g P_2O_5 kg^{-1}. On the other hand, apatite crystallizes over a larger SiO_2 range in magmas with a higher P content than this level (Green and Watson 1982). As the total P content of the basaltic andesite and basaltic tephras used in the present study was 2.8 g P_2O_5 kg^{-1} or less, crystallization of apatite could not occur and P was considered to be distributed to other minerals.

Dissolution of apatite is dependent on pH (Fig. 6.31d). Two apatite samples of sedimentary origin (Florida and Makatea) were treated with 10 mmol L^{-1} of citrate or dilute HCl. The amount of P dissolved from the two apatite samples increased with decreasing final pH values and was larger in the citrate treatment than in the dilute HCl treatment, especially in the pH range of 5–6. This is possibly due to chelation of Ca and donation of H^+ by citrate.

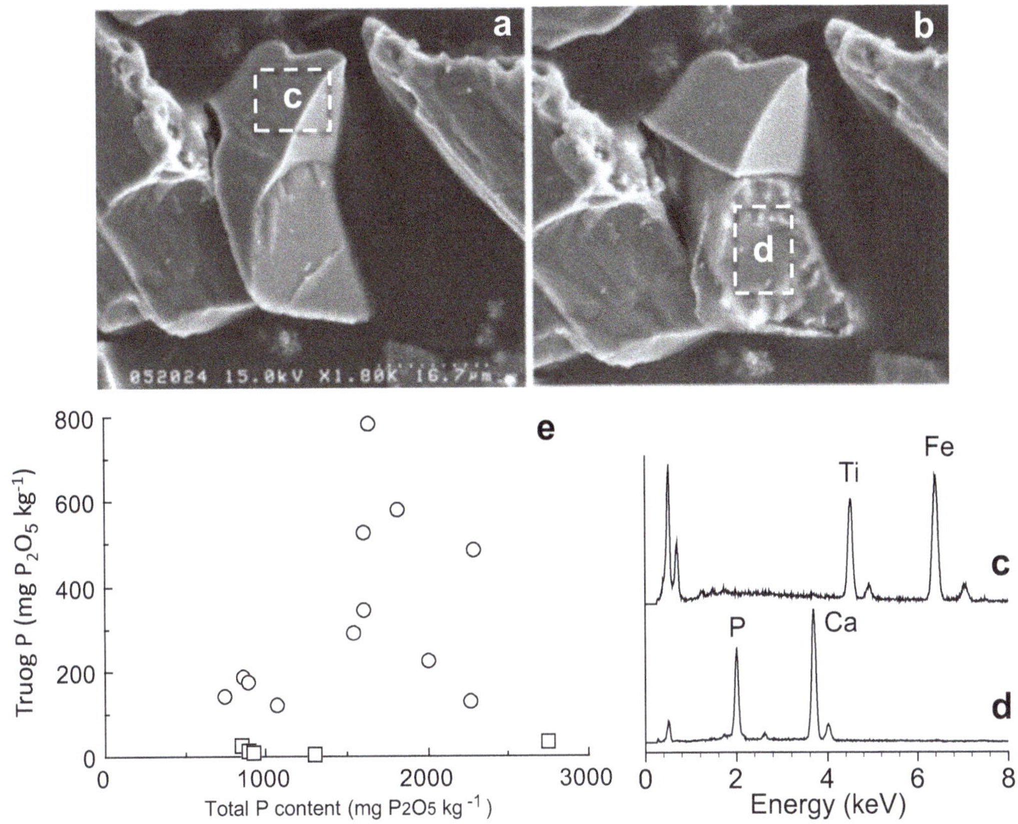

Fig. 6.30 Dissolution of apatite in the Truog extracting solution. (**a**) Magnified SEM image of particles included Fig. 6.29c, (**b**) SEM image after the treatment with Truog solution, (**c** and **d**) EDX spectra of the dashed areas (**c**) and (**d**) of Fig. 6.30a, b, respectively, (**e**) relationship between total P content and Truog P of rhyolitic to andesitic (open circles) and basaltic-andesitic to basaltic (open squares) tephras

In contrast, when the two apatite samples were treated with 10 mmol L^{-1} oxalate, the amount of P dissolved was much smaller than those of the treatment with citrate or HCl in the final pH range between 3.4 and 4.3 (Fig. 6.31d). When the final pH was about 3 with the treatment by citrate or dilute HCl, the amount of P dissolved was more than 96% of the original P content of the samples. When the final pH was 3.4 with the treatment by oxalate, the amount of P dissolved was only about 10% of the original P content and was not much different from those at final pH of 6–7. The low P dissolution with oxalate treatment, compared with the dissolution of citrate or HCl treatment, was close to the result using the Pinatubo tephra containing apatite of igneous origin. The formation of a Ca oxalate coating on the surface of apatite from the oxalate treatment of the Florida apatite was revealed by XRD (Nanzyo et al. 1999). Similar results were also obtained for the Makatea apatite.

The distribution of the Ca oxalate by the oxalate treatment of the apatite samples was elucidated by SEM EDX analysis. Figure 6.31a shows an SEM image of the surface (right-hand side) and the section (left-hand side) of an oxalate-treated Florida apatite particle at pH4. According to the selected area analysis, the peak intensity for P-Kα of the surface of the particle (dashed square (b) in Fig. 6.31a) was much

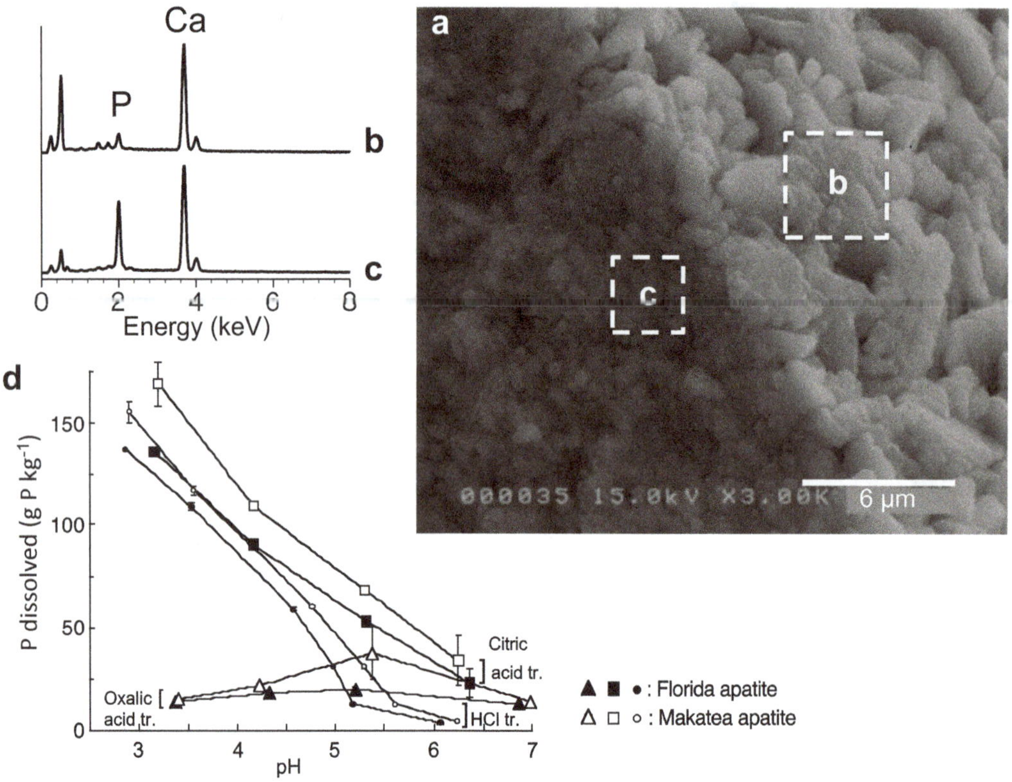

Fig. 6.31 Formation of Ca oxalate coating on an apatite particle. (**a**) SEM image of surface (**b**) and cross section (**c**) of Florida apatite treated with oxalate (pH 4), (**b** and **c**) EDX spectra of dashed squares (**b**) and (**c**) of Fig. 6.31a, (**d**) changes in P dissolution with pH when treated with oxalate, citrate, and HCl

weaker than that of the section (dashed square (c) in Fig. 6.31a). The thickness of the Ca oxalate coating was estimated to be less than 2 μm based on the results from the line scan for P. Thus, it was confirmed that Ca oxalate was formed as a surface coating material by oxalate treatment of the apatites and that the Ca oxalate coating inhibited further dissolution of apatite (Nanzyo et al. 1999).

Chickpea roots secrete organic acids (Ohwaki and Hirata 1992). A decrease in pH value and an increase in P concentration was detectable when chickpea was cultivated in the Pinatubo volcanic ash with a solution containing no P (Nakamaru et al. 2000).

As shown in Figs. 6.29 and 6.30, significant amounts of P are contained in fresh tephra, and the major P-bearing mineral in the andesitic to rhyolitic tephra is apatite. During the process of Andisol formation, apatite is dissolved gradually and the dissolved P is sorbed by active Al and Fe. Figure 6.32a shows an allophanic Andisol profile at Yunodai, Aomori Prefecture, Japan. The C horizon has 0.14% of oxalate-extractable Al (Al_o) and 1.7% of oxalate-extractable Fe (Fe_o), indicating that it is weakly weathered. Figure 6.32b shows an SEM image of weathered apatite (in and around the dashed square Fig. 6.32c) and allophanic material (in and around the dashed square Fig. 6.32d), determined from EDX spectra (c) and (d), respectively, in

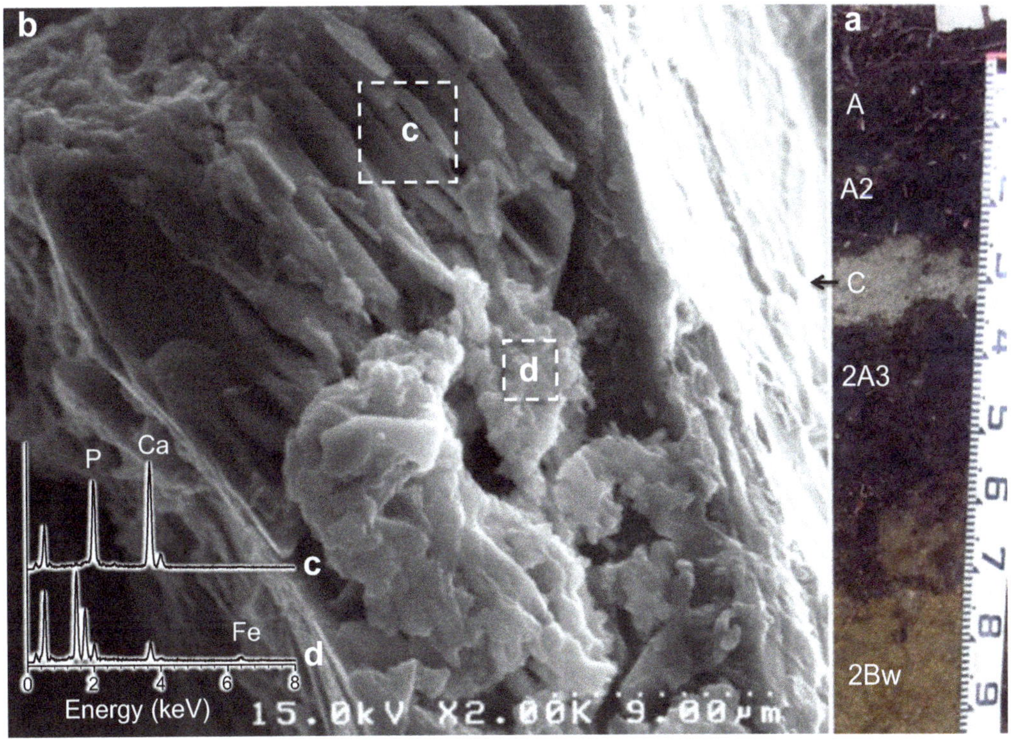

Fig. 6.32 Weathering of apatite in an Andisol. (**a**) Soil profile of Hapludand, (**b**) SEM image of a sand particle including partially weathered apatite and allophanic material, (**c** and **d**) EDX spectra of the dashed areas (**c**) and (**d**), respectively

Fig. 6.32b. Parallel-oriented thin prismatic form is a commonly found character of weathered apatite in a young Andisol. As the EDX spectrum Fig. 6.32c lacks X-ray counts for Al and Fe, there is no coating of active Al or Fe materials on the surface of the weathered apatite, suggesting that weathering of apatite in and around the dashed square (c) is congruent dissolution of apatite. The reaction of apatite and active Al is different from the formation of a Ca oxalate coating on the surface of apatite (Fig. 6.31).

6.4.2 Reactions of Phosphate with Active Al and Fe Materials

Phosphate concentration in soil water is generally low, and it is less than 10^{-6} mol L^{-1} in mature Andisols, which are rich in active Al materials. This low P concentration in soil water is less than that at dissolution equilibrium of $AlPO_4 \cdot 2H_2O$, variscite. The reason for the low P concentration in the soil water is that P is sorbed by active Al and Fe. In general, P sorption by active Al and Fe occurs at lower P concentration than precipitation of $AlPO_4 \cdot 2H_2O$ from soil water. With increase in the amount of P sorption, P concentration in soil water and the amount of plant-available P increase gradually.

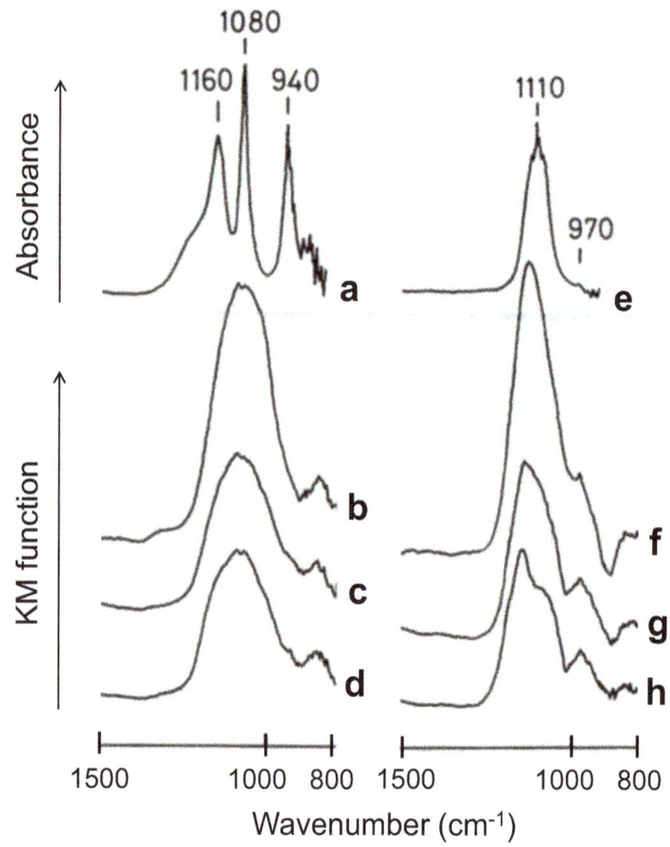

Fig. 6.33 Changes in infrared absorption spectra of phosphate and sulfate with sorption by aluminum hydroxide gel. (**a**) NaH_2PO_4 in water, (**b**) sorbed $H_2PO_4^-$ at room temperature and atmospheric pressure, (**c**) sorbed $H_2PO_4^-$ at room temperature and reduced pressure, (**d**) sorbed $H_2PO_4^-$ under reduced pressure at 373 K, (**e**) Na_2SO_4 in water, (**f**) sorbed SO_4^{2-} at room temperature and atmospheric pressure, (**g**) sorbed SO_4^{2-} at room temperature and reduced pressure, (**h**) sorbed SO_4^{2-} under reduced pressure at 373 K

As a P sorption product is not detectable by XRD, infrared absorption (IR) spectroscopy is effective. Parfitt et al. (1976), Nanzyo and Watanabe (1982), and Nanzyo (1986) used IR spectroscopy to examine P sorption by active Fe materials. In IR spectroscopy, the vibration mode of a molecule, of which the dipole moment changes, is detected. Figure 6.33a is an infrared absorption spectrum of dihydrogen phosphate in water. There are three or more absorption bands related to P–O stretching vibration. A distinctive change in the IR spectrum occurred when P was sorbed by the Al hydroxide gel (Fig. 6.33b). In order to highlight the IR spectrum of sorbed P, a differential method was used in the IR spectroscopy of Fig. 6.33b, c, d, f, g, h. The absorption bands of phosphate were converted to a broad one (Fig. 6.33b) after sorption by the Al hydroxide gel, suggesting that the product is a material close to noncrystalline aluminum phosphate. The simple and broad IR absorption band changed little under reduced pressure and heating (Fig. 6.33c, d). However, the absorption band is broader than that for noncrystalline aluminum phosphate. The sorbed P by Al hydroxide gel may have an intermediate property between surface complex and noncrystalline Al phosphate (Nanzyo 1984, 1987, 1988). The IR spectrum of the P accumulated in Andisols as a result of application of P fertilizer is similar to those for Fig. 6.33b, c, d (Nanzyo 1987).

An inverse change is observable for sulfate in water and sorbed sulfate by Al hydroxide gel. Sulfate shows a rather simple IR absorption band in water (Fig. 6.33e) due to the high symmetry of SO_4^{2-} (Fig. 6.33e), whereas the IR bands of sorbed

Fig. 6.34 Ammonium taranakite formed from the reaction of an Andisol and ammonium dihydrogen phosphate. (**a**) Optical micrograph, (**b**) SEM image, (**c**) EDX spectrum, (**d**) XRD pattern of (**a**), (**e**) reference XRD pattern (Lehr et al. 1967)

sulfate by the Al hydroxide gel have an increased degree of splitting with reduced pressure and heating. The symmetry of SO_4^{2-} was reduced with an increase in the interaction between sulfate and the Al hydroxide gel surface under reduced pressure and heating. As a result, the IR absorption bands of sulfate split (Figs. 6.33f, g, h). The contrasting IR spectral changes between P and sulfate when sorbed by Al hydroxide gel is believed to reflect the difference in their reactions with the Al hydroxide gel.

Taranakite [$Al_5(NH_4)_3H_6(PO_4)_8 \cdot 18H_2O$] is the crystalline product from reactions between P and soil inorganic constituents (Fig. 6.34) (Wada 1959). Allophanic clays, gibbsite, halloysite, and Andisols can be a source of Al for taranakite. Figure 6.34a shows an optical micrograph of ammonium taranakite formed from the reaction of an allophanic Andisol and $NH_4H_2PO_4$. To obtain the taranakite particles, the less than 38 μm fraction of the Andisol was used for the reaction and the taranakite particles were then separated using a 53 μm sieve. The light-brown color of one particle is due to adhesion of fine soil particles. In the EDX spectrum (Fig. 6.34c), the absence of N is due to the property of the window material of the X-ray detector. The powder XRD pattern (Fig. 6.34d) is identical to that of the reference (Fig. 6.34e). A potassium-substituted phase, [$Al_5K_3H_6(PO_4)_8 \cdot 18H_2O$], also exists. An $NH_4H_2PO_4$ concentration higher than 0.2 mol L^{-1} and an acidic pH of around 4 are needed to form taranakite from Andisols. However, even from cultivated Andisol with high P accumulation, taranakite is not formed by the addition of 1 mol L^{-1} NH_4Cl and pH adjustment.

6.4.3 *Struvite*

Struvite ($MgNH_4PO_4 \cdot 6H_2O$) is a major crystalline phosphate in compost (Komiyama et al. 2013). Swine manure compost and chicken manure compost have a total P_2O_5 content of 4–5%. Cattle manure compost has a lower amount. Struvite plays an important role in recycling P in the soil–plant system.

Figure 6.35a shows pelletized swine manure compost. According to the optical micrograph (Fig. 6.35b), the struvite is a colorless transparent crystal. Figure 6.35c shows an SEM image including the same struvite particle and indicating that the major axis of the crystal is approximately 120 μm. The dashed square d in Fig. 6.35c includes Mg and P, supporting the conclusion that this crystal is struvite, although N is lacking in the EDX spectrum due to the property of the window material of the

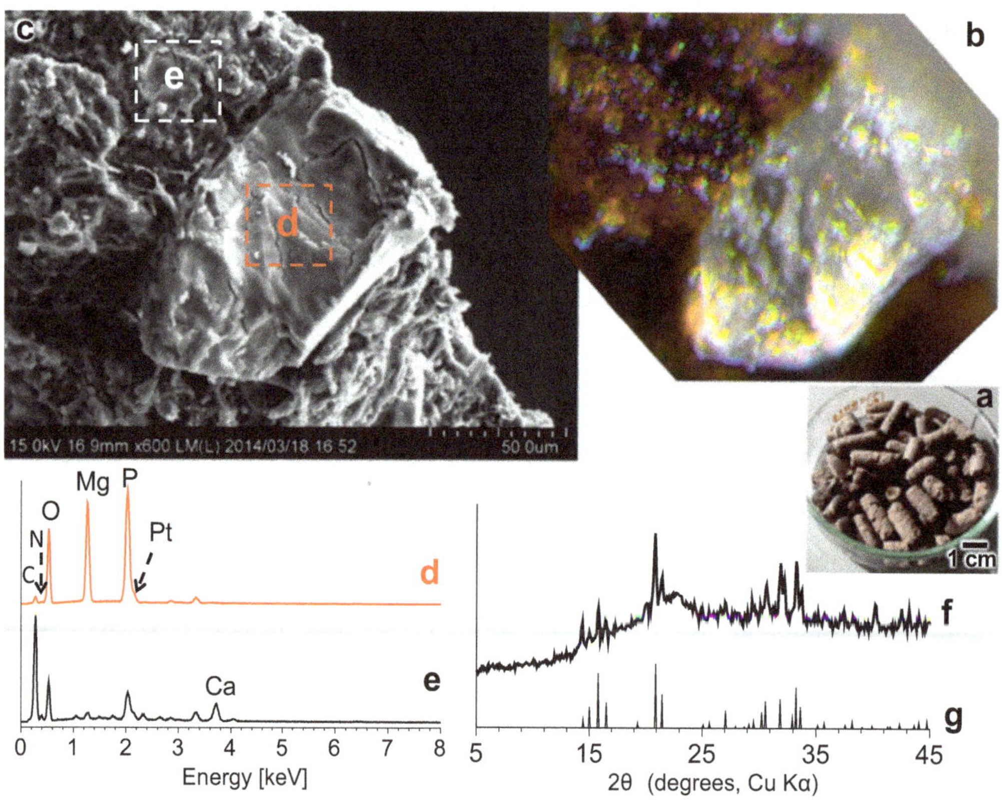

Fig. 6.35 Struvite in pelletized swine manure compost. (**a**) Fermented and pelletized swine manure (FPSM), (**b**) optical micrograph of struvite formed in the FPSM, (**c**) SEM image of a piece of FPSM, (**d** and **e**) EDX spectra of dashed areas (**d**) and (**e**) in (**c**), respectively, (**f** and **g**) XRD patterns of powdered FPSM and reference struvite (Lehr et al. 1967), respectively

X-ray detector. The bulk area of the pelletized compost, the other dashed square e of Fig. 6.35c, shows an EDX spectrum with very small peaks for Mg, P, K, and Ca. A power XRD pattern obtained from the ground sample (Fig. 6.35f) includes XRD peaks identical to those of the reference struvite (Fig. 6.35g).

6.4.4 *Phosphorus Management in Farmlands*

For sustainable phosphorus management in farmlands, measurement and evaluation of the plant-available P level in soil and the P recovery by crop plants is needed. Syers et al. (2008) reviewed several methods for estimating the recovery of phosphorus fertilizers (Fig. 6.36) and stated that the balance method should be re-evaluated despite the frequent use of the difference method. A shortcoming of the difference method is that soils with high residual P show low recovery percentages whereas the P uptake by crop plants is not low. In contrast, the P recovery by the balance method shows high percentages for soils with high residual P.

The plant-available P level in the world's farmlands may be diverse, depending on history of the soil and the fertilizer management. If all of the total P uptake, crop yield, and crop quality are adequate, and the P recovery rate is 100% or higher by the balance method, P application should be adjusted to minimize P problems, such as fertilizer cost, environmental impact, and dissipation of P resources. Many paddy

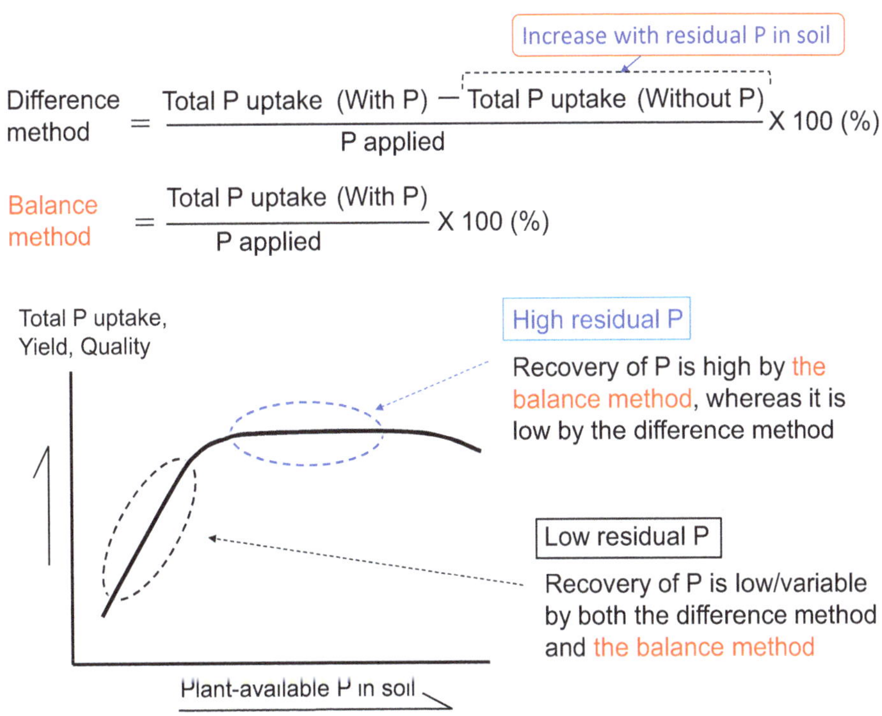

Fig. 6.36 Methods for estimating the recovery of phosphorus fertilizers. (Prepared after Syers et al. 2008)

field soils in Japan are at this P level, although the plant-available P level increases under reducing conditions (Figs. 5.14 and 5.21). The active Al content in most paddy field soils in Japan is much lower than those of mature Andisols. P management at a P recovery rate of around 100% by the balance method can be sustainable.

For farmlands with low plant-available P and residual P, P recovery is low or variable by both the balance and difference methods. The P recovery rate and the relation between plant-available P in soil and total P uptake may vary with soil properties and managements, properties of P fertilizers, application methods of P fertilizers, properties of crop plants, and climatic conditions. P application methods designed to improve P recovery are needed.

In uncultivated Andisols, the P concentration of soil water is so dilute that wild plants concentrate P by an order of 10^3 (Lyu et al. 2016). Many plants obtain dilute P from soil through a symbiosis with arbuscular mycorrhiza. On the other hand, *Brassica* plant roots develop as if they were foraging P in Andisol with low plant-available P (Nanzyo et al. 2002, 2004). Pelletized chicken or swine manure compost (Fig. 6.35a) is suitable P fertilizer for the phosphorus-foraging root growth of the Japanese radish in this case. Growth of the Japanese radish is normal, and the P recovery rate is high. *Brassica* plants are not symbiotic with arbuscular mycorrhiza.

Further readings for phosphorus in soil are Khasawneh et al. (1980), and Sims and Shapley (2005).

References

Adachi K, Kajimo M, Zaizen Y, Igarashi Y (2013) Emission of spherical cesium-bearing particles from an earyly stage of the Fukushima nuclear accident. Sci Rep 3:2554

Agus F, Tinning G (eds) (2008) International Workshop on Post Tsunami Soil Management, Proceedings 'Lessons learned for agriculture and environmental restoration in the aftermath of the 2004 tsunami, Bogor, Indonesia, 1–2 July 2008, pp 1–177

Amundson R, Berhe AA, Hopmans JW, Olson C, Sztein AE, Sparks DL (2015) Soil and human security in the 21st century. Science 348:6235

Doner HE, Lynn CW (1989) Carbonate, halide, sulfate, and sulfide minerals. In: Dixon JB, Weed SB (eds) Minerals in soil environments, 2nd edn. SSSA, Madison, pp 279–330

Edixhoven JD, Gupta J, Savenije HHG (2014) Recent revisions of phosphate rock reserves and resources: a critique. Earth Syst Dynam 5:491–507

Endo T, Yamamoto S, Honna T, Eneji AE (2002) Sodium-calcium exchange selectivity as influenced by clay minerals and composition. Soil Sci 167:117–125

Ericksen J (2009) Soil sulfur cycling in temperate agricultural systems. Adv Agron 102:55–89

Evangelow VP, Phillips RE (2005) Cation exchange in soils. In: Chemical processes in soils, SSSA book series, no.8. SSSA, Madison, pp 343–409

Evrard O, Chartin C, Onda Y, Lepage H, Cerdan O, Lefevre I, Ayrault S (2014) Renewed soil erosion and remobilization of radioactive sediment in Fukushima coastal rivers after the 2013 typhoons. Sci Rep 4:4574

Geospacial Information Authority of Japan (2014) Extension site of distribution map of radiation dose, etc. http://ramap.jmc.or.jp/map/eng/

Green TH, Watson EB (1982) Crystallization of apatite in natural magmas under high pressure, hydrous conditions, with particular reference to 'orogenic' rock series. Contrib Mineral Petrol 79:96–105

Hirose K (2016) Fukushima Daiichi nuclear plant accident: atomospheric and oceanic impacts over the five years. J Environ Radioact 157:113–130

Inao E, Kamiyama K, Moriya K, Konno C, Onodera K, Shima H, Ito T, Kanno H (2013) Chemical property of tsunami sediment deposited following the East Japan great earthquake (south part of Miyagi Prefecture). Agric Hortic Res Inst Res Rep 81:63–87 (In Japanese with English summary)

Ippolito JA, Barbarick KA, Elliott HA (2011) Drinking water treatment residuals: a review of recent uses. J Environ Qual 40:1–12

Kamphorst A, Bolt GH (1976) Saline and sodic soils. In: Bolt GH, Bruggenwert MGM (eds) Soil chemistry A. Basic elements, developments in soil science 5A. Elsevier, Amsterdam/Oxford/New York, pp 171–191

Kanno H (2017) Impact of the 2011 Tohoku-oki earthquake tsunami on cultivated oils in Miyagi Prefecture, northeastern Japan: an overview. In: Santiago-Fandino V et al (eds) The 2011 Japan earthquake and tsunami: reconstruction and resortation, Advances in natural and technological hazards research, vol 47. Springer, Dordrecht, pp 341–354

Kanno H, Nanzyo M, Takahashi T (2013) Tour guide of 60th field trip of Japanese Society of Pedology, pp 62–73

Khasawneh FE, Sample EC, Kamprath EJ (1980) The role of phosphorus in agriculture. ASA-CSSA-SSSA, Madison

Komiyama T, Niizuma S, Fujisawa E, Morikuni (2013) Phosphorus compounds and their solubility in swine manure compost. Soil Sci Plant Nutr 59:419–426

Kopittke PM, So HB, Menzies NW (2006) Effect of ionic strength and clay mineralogy on Na-ca exchange and the SAR-ESP relationship. Eur J Soil Sci 57:626–633

Lehr JR, Brown EH, Frazier AW, Smith JP, Thrasher RD (1967) Crystallographic properties of fertilizer compounds. Chemical engineering bulletin 6. Tennessee Valley Authority, Knoxville, TN

Lyu Y, Tang H, Li H, Zhang F, Rengel Z, Whalley WR, Shen J (2016) Major crop species show differential balance between root morphological and physiological responses to variable phosphorus supply. Front Plant Sci 7:1939

McBride MB (1989) Surface chemistry of soil minerals. In: Dixon JB, Weed SB (eds) Minerals in soil environments, SSSA book series no.1. SSSA, Madison, pp 35–88

Miller WP, Frenkel H, Newman KD (1990) Flocculation concentration and sodium-calcium exchange of kaolinitic soil clays. Soil Sci Soc Am J 54:346–351

Ministry of Agriculture, Forestry and Fisheries Japan (2012) FY2011 annual report on food, agriculture and rural areas in Japan, Summary

Minoura K, Imamura F, Sugawara D, Kono Y, Iwashita T (2001) The 869 Jogan tsunami deposit and recurrence interval of large-scale tsunami on the Pacific coast of Northeast Japan. J Nat Disaster Sci 23:83–88

Monecke K, Finger W, Klarer D, Kongko W, McAdoo BG, Moore AL, Sudrajat SU (2008) A 1,000-year sediment record of tsunami recurrence in northern Sumatra. Nature 455:1232–1234

Mukai H, Hirose A, Motai S, Kikuchi R, Tanoi K, Nakanishi T, Yaita T, Kogure T (2016) Cesium adsorption/desorption behavior of clay minerals considering actual contamination conditions in Fukushima. Sci Rep 6:21543

Nakamaru Y, Nanzyo M, Yamasaki S (2000) Utilization of apatite in fresh volcanic ash by Pigeonpea and chickpea. Soil Sci Plant Nutr 46:591–600

Nakanishi T, Tanoi K (eds) (2013) Agicultural implicaitons of the Fukushima nuclear accident. Springer, Tokyo

Nakanishi T, Tanoi K (eds) (2016) Agicultural implicaitons of the Fukushima nuclear accident: the first three years. Springer, Tokyo

Nakao A, Ogasawara S, Sano O, Ito T, Yanai J (2014) Radiocesium sorption in relation to clay mineralogy of paddy soils in Fukushima, Japan. Sci Total Environ 468:523–529

Nakaya T, Tanji H, Kiri H, Hamada H (2010) Developing a salt-removal plan to remedy Tsunami-caused salinity damage to farmlands: Case study for an area in Southern Thailand. JARQ, 44: 159–165

Nanzyo M (1984) Diffuse reflectance infrared spectra of phosphate sorbed on alumina gel. J Soil Sci 35:63–69

Nanzyo M (1986) Infrared spectra of phosphate sorbed on iron hydroxide gel and the sorption products. Soil Sci Plant Nutr 32:51–58

Nanzyo M (1987) Formation of noncrystalline aluminum phosphate through phosphate sorption on allophanic ando soils. Commun Soil Sci Plant Anal 18:735–742

Nanzyo M (1988) Phosphate sorption on the clay fraction of Kanuma pumice. Clay Sci 7:89–86

Nanzyo M (2012) Impacts of tsunami (March 11, 2011) on paddy field soils in Miyagi Prefecture, Japan. J Integr Field Sci 9:3–10

Nanzyo M, Watanabe Y (1982) Diffuse reflectance infrared spectra and ion-adsorption properties of the phosphate surface complex on goethite. Soil Sci Plant Nutr 28:359–368

Nanzyo M, Yamasaki S (1998) Phosphorus bearing mineral in fresh, andesite and rhyolite tephras in northern part of Japan. Phosphorus Res Bull 8:95 100

Nanzyo M, Takahashi T, Sato A, Shoji S, Yamada I (1997) Dilute acid-soluble phosphorus in fresh air-borne tephras and fixation with an increase in active aluminum and iron. Soil Sci Plant Nutr 43:839–848

Nanzyo M, Nakamaru Y, Yamasaki S (1999) Inhhibition of apatite dissolution due to formation of calcium o xalate coating. Phosphorus Res Bull 9:17–22

Nanzyo M, Shibata Y, Wada N (2002) Complete contact of *Brassica* roots with phosphates in a phosphorus-deficient soil soil Sci. Plant Nutr 48(6):847–853

Nanzyo M, Ebuchi Y, Kanno H (2003) Apatite in the pyroclastic flow deposits (1990–1995) of the Unzen volcano, Japan, and its utilization by buckwheat. Phosphorus Res Bull 16:1–10

Nanzyo M, Wada N, Kanno H (2004) Phosphorus foraging root development of *Brassica rapa* nothovar. grown in a phosphorus-deficient nonallophanic. Andisol Plant Soil 265:325–333

Nanzyo M, Kanno H, Takeda A (2014) Vertical distribution of radiocesium in side bar deposits of the Utsushi and Agano rivers, Japan. Clay Sci 18:43–52

Nanzyo M, Takeda A, Hio A, Ito K, Kanno H, Takahashi T (2015) Vertical distribution of radocaesium in tsunami-affected farmland soils in Miyagi Prefecture, Japan. Icobte 2015 Fukuoka abstract book, 13th international conference on the biogeochemistry of trace elements, July 12–16, 2015, Fukuoka International Congress Center, Fukuoka, Japan

Niigata City (2014) Radioactive elements in water treatment residual generated in Aaganogawa water purification plant. http://www.city.niigata.lg.jp/kurashi/jyogesuido/suido/sinsai/tuti_chosa/aganogawa.html

Norio O, Ye T, Kajitani Y, Shi PJ, Tatano H (2011) The 2011 eastern Japan great earthquake disaster: overview and comments. Int J Disaster Risk Sci 2:34–42

Oda K, Miwa E, Iwamoto A (1987) Compact database. Jpn J Soil Sci Plant Nutr 61:112–131 (In Japanese)

Ohwaki Y, Hirata H (1992) Differences in carboxylic acid exudation among P-starved leguminous crops in relation to carboxylic acid contents in plant tissues and phospholipid level in roots. Soil Sci Plant Nutr 38:235–243

Okal EA (2011) Tsunamigenic earthquakes: past and present milestones. Pure Appl Geophys 168:969–995

Parfitt RL, Russell JD, Farmer VC (1976) Confirmation of the surface structures of goethite (α-FeOOH) and phosphate goethite by infrared spectroscopy. J Chem Soc, Faraday I 72:1082–1087

Saitama Prefecture (2014) Radioactivity of water treatment residual generated in water purification plant. http://www.pref.saitama.lg.jp/page/hasseido-sokuteikekka.html

Saito K, Shimbori T, Draxler R (2015) JAM's regional atmospheric transport model calculations for the WMO technical task team on meteorological analyses for Fukushima Daiichi nuclear power plant accident. J Environ Radioact 139:185–199

Satake K, Rabinovich A, Kanoglu U, Tinti T (2011) Introduction to "tsunamis in the World Ocean: past, present, and future. Volume I". Pure Appl Geophys 168:963–968

Shainberg I, Oster JD, Wood JD (1980) Sodium=calcium exchange in montmorillonite and illite suspensions. Soil Sci Soc Am J 44:960–964

Shima H, Onodera K, Kanazawa Y, Satoh K, Onodera H, Abe T, Wakashima A, Inao E, Moriya K, Konno C, Kamiyama K, Itoh T, Kanno H (2012) Chemical property of tsunami sediment deposited following the East Japan great earthquake (northern part of Miyagi Prefecture). Bull Miyagi Prefectural Furukawa Agric Exp Station 10:33–42 (In Japanese with English summary)

Shiozawa S (2013) Vertical migration of radiocesium fallout in soil in Fukushima. In: Nakanishi TM, Tanoi K (eds) Agricultural implications of the Fukushima nuclear accident. Springer, Tokyo/Heiderberg/New York/Dordrecht/London, pp 49–60

Sims JT, Sharpley AN (2005) Phosphorus: agriculture and the environment. No.46 in the series Agronomy. ASA-CSSA-SSSA, Madison

Srisutam C, Wagner J-F (2010) Tsunami sediment characteristics at the Thai Andaman Coast. Pure Appl Geophys 167:215–232

Steinhauser G, Brandl A, Johnson TE (2014) Comparison of the Chernobyl and Fukushima nuclear accidents: a review of the environmental impacts. Sci Total Envron 470–471:800–817

Syers K, Johnston E, Curtin D (2008) Efficiency of soil and fertilizer phosphorus: reconciling changing concepts of soil phosphorus behavior with agronomic information, FAO fertilizer and plant nutrition bulletin 18. FAO, Rome

Szczuciński W, Kokociński M, Rzeszewski M, Chague-Goff C, Cachao M, Goto K, Sugawara D (2012) Sediment sources and sedimentation processes of 2011 Tohoku-oki tsunami deposits on the Sendai Plain, Japan—Insights from diatoms, nannoliths and grain size distribution. Sediment Geol 282:40–56

Thomas G (1982) Exchangeable cations. In: Methods of soil analysis part 2, Chemical and microbiological properties, 2nd edn. ASA-SSSA, Madison, pp 159–165

Tsukada H, Takeda A, Hisamatsu S, Inaba J (2008) Concentration and specific activity of fallout ^{137}Cs in extracted and particle-size fractions of cultivated soils. J Environ Radioact 99:875–881

U.S. Salinity Laboratory Staff (1954) Diagnosis and improvement of saline and alkali soils. In: USDA agriculture handbook 60. U.S. Government Printing Office, Washington, DC, pp 25–30

Vorob'eva LA, Pankova EI (2008) Saline-alkali soils of Russia. Eurasian Soil Sci 41:517–532

Wada K (1959) Reaction of phosphate with allophane and halloysite. Soil Sci 87:325–330

Yomiuri News Paper (2011) Cesium of water treatment residual generated from Aganogawa water purification plant. https://database.yomiuri.co.jp/ rekishikan/ (August, 2014)

Index

M. Nanzyo, H. Kanno, *Inorganic Constituents in Soil*,
https://doi.org/10.1007/978-981-13-1214-4